Industrial Complex Analysis and Regional Development

TECHNOLOGY PRESS BOOKS

Regional Science Studies *edited by Walter Isard*

LOCATION AND SPACE-ECONOMY
  By Walter Isard
THE LOCATION OF THE SYNTHETIC-FIBER INDUSTRY
  By Joseph Airov
INDUSTRIAL COMPLEX ANALYSIS AND REGIONAL DEVELOPMENT
  By Walter Isard, Eugene W. Schooler, and Thomas Vietorisz

Other Books in the Social Sciences

THE INFLATIONARY SPIRAL: THE EXPERIENCE IN CHINA, 1939–1950
  By Chang Kia-Ngau
THE TAO OF SCIENCE: AN ESSAY ON WESTERN KNOWLEDGE AND EASTERN WISDOM
  By R. G. H. Siu
SOVIET EDUCATION FOR SCIENCE AND TECHNOLOGY
  By Alexander G. Korol
THE ECONOMICS OF COMMUNIST EASTERN EUROPE
  By Nicolas Spulber
ON HUMAN COMMUNICATION: A REVIEW, A SURVEY, AND A CRITICISM
  By Colin Cherry
SCIENCE AND ECONOMIC DEVELOPMENT: NEW PATTERNS OF LIVING
  By Richard L. Meier
MOSCOW AND THE COMMUNIST PARTY OF INDIA
  By John H. Kautsky
LANGUAGE, THOUGHT, AND REALITY
  By Benjamin Lee Whorf
  Edited by John B. Carroll
THE TERMS OF TRADE: A EUROPEAN CASE STUDY
  By Charles P. Kindleberger
MACHINE TRANSLATION OF LANGUAGES
  Edited by W. N. Locke and A. D. Booth
NINE SOVIET PORTRAITS
  By Raymond A. Bauer
THE PROSPECTS FOR COMMUNIST CHINA
  By W. W. Rostow and others
LABOR MOBILITY AND ECONOMIC OPPORTUNITY
  By Members of the Social Science Research Council

NATIONALISM AND SOCIAL COMMUNICATION
  By Karl W. Deutsch
INDUSTRIAL RELATIONS IN SWEDEN
  By Charles A. Myers
MID-CENTURY: THE SOCIAL IMPLICATIONS OF SCIENTIFIC PROGRESS
  Edited by John F. Burchard
CYBERNETICS: OR CONTROL AND COMMUNICATION IN THE ANIMAL AND THE MACHINE
  By Norbert Wiener

Industrial Complex Analysis and Regional Development:

A Case Study of Refinery–Petrochemical–Synthetic-Fiber Complexes and Puerto Rico

by
WALTER ISARD
University of Pennsylvania

EUGENE W. SCHOOLER
University of Pennsylvania

THOMAS VIETORISZ
United Nations
Economic Commission for Latin America

PUBLISHED JOINTLY BY
The Technology Press of
The Massachusetts Institute of Technology
AND
John Wiley & Sons, Inc., New York
Chapman & Hall, Limited, London

Copyright © 1959

by

The Massachusetts Institute of Technology

All rights reserved. This book or any part thereof must not be reproduced in any form without the written permission of the publisher.

Library of Congress Catalog Card Number: 59-13032

Printed in the United States of America

Preface

This book is the third in a series of Regional Science Studies. In line with the objectives of this series, it attempts to fill an important gap in the existing stock of analytical techniques in the fields of regional science, industrial geography and location, and regional economics and planning. It develops the industrial complex approach to analysis, a new approach which aims to complement and cultivate the area lying between input–output and linear programming methods on the one hand, and individual industry comparative-cost study on the other. This approach is fully developed in the form of a case study involving Puerto Rico as the primary region, and oil refining, petrochemical, synthetic fiber, and fertilizer processes as the chief production activities.

As with the preceding volume in the Regional Science Studies series, this third book develops empirical materials which should be of interest to many. The detailed data, analyses, and results should prove valuable not only to companies and personnel engaged in oil refining, and in the production of petrochemicals, fertilizers, and synthetic fibers, but also to firms and business officials in various fields who seek diversification and new, profitable channels for expansion. Also, the approach as well as the empirical materials should be of considerable significance to government agencies, business units, planners, and scholars concerned with economic development. The materials relate to industrial growth in an underdeveloped region (Puerto Rico). The approach demonstrates how study of development potentials in terms of individual industry investigations may be supplemented and complemented with analysis of the interrelations of these industries—without being involved in general frameworks such as input–output, which too frequently are nonoperational for underdeveloped and other regions.

It must be strongly stated that, because this book is primarily designed to be a case study, it does not attempt to appraise generally the industrial complex technique as an analytical approach. A full appraisal and a general formulation of this technique, partly based on this case study, are contained in Walter Isard *et al., Methods of Regional Analysis,* forthcoming.

Much of the basic empirical information which forms the foundation of the present study was developed by the authors and their associates in connection with other individual projects. Specifically, heavy reliance is placed on studies by J. Robert Lindsay, Jr., and Joseph Airov for data on oil refinery activities and synthetic fiber activities, respectively.

We are grateful to many industrial and research firms and their officials for generous assistance in the collection and evaluation of technical information. Among these firms are Stone & Webster Engineering Corporation, the Lummus Company, Arthur D. Little Company, Badger Manufacturing Company, M. W. Kellogg Company, the National Research Corporation, Gulf Research and Development Company, Phillips Petroleum Company, Standard Oil Company of Indiana, Cosden Petroleum Corporation, Chemstrand Corporation, Tennessee Eastman Company, Union Carbide Corporation, E. I. du Pont de Nemours & Company. The Oklahoma Planning and Resources Board, Industrial Development Division, very kindly made available a report, prepared by the Blaw-Knox Company, on the production of ethylene and its derivatives. We wish to express our appreciation also to Mr. John F. O'Donnell and to Professors R. F. Baddour, C. N. Satterfield, and C. P. Kindleberger of M. I. T. for supplying information and helping to guide judgment on various matters of a technical nature. The U.S. Department of Commerce, and particularly officials of its Office of Technical Services and Area Development Division, have been of assistance in several ways.

Finally, we are deeply indebted to the Social Science Research Center, University of Puerto Rico, for its generous support of this project, and in particular to its director, Millard Hansen, for his patient administration and constant encouragement. Substantial additional support of this research, especially in its conceptual stages, was provided by a grant from Resources for the Future, Inc.

Philadelphia
September, 1959

WALTER ISARD
EUGENE W. SCHOOLER
THOMAS VIETORISZ

Contents

CHAPTER	PAGE
Summary	1

**1 • Possible Approaches to the Analysis
of Puerto Rico as a Region** 5
 Introduction   5
 The Broad Economic Development Approach   7
 Comparative-Cost Analysis for Individual Industries   7
 The Location Quotient   9
 The Coefficient of Localization   10
 The Labor Coefficient   11
 The Interregional Input–Output Approach   12
 Some Reasons for the Choice of an Industrial
 Complex Approach   25

2 • The Choice of Relevant Industrial Complexes 27
 The Resource Setting   27
 Industrial Activities Selected   28
 Activities Excluded   32
 Type of Complex to Be Evaluated   33
 Basic Postulates. Plan of the Study   36

**3 • The Individual Production Activities:
Inputs and Outputs** 39
 The Basic Activity Matrix   39
 Commodities Omitted from the Basic Matrix   50
 Capital Inputs   52
 Direct Labor Inputs   56
 Indirect Costs Based on Capital and Direct Labor Inputs   58

CONTENTS

CHAPTER | PAGE

4 • The Full Programs ... 62
Program Restrictions 62
The General Program Pattern 68
Detailed Presentation of Sample Programs 71
   Flowsheets and Subprograms 71
   Production Programs 79
   Total Inputs and Outputs, by Program 88

5 • Derivation of Cost and Revenue Differentials, by Commodity ... 95
Crude Oil and Liquid Refinery Products 96
Gaseous Fuel Products 103
Fuel Inputs: Steam and Power 104
LPG and Coke 107
Liquid, Solid, and Gaseous Chemicals 109
Commodities Subject to Special Consideration 114
Staple and Continuous-Filament Fibers 116
Fertilizer 118
Miscellaneous (Zero-Differential) Commodities 119
Labor 120
Adjustments for Different Raw-Material Sources 121
Adjustments because of Split-Location Pattern
on Mainland 122

6 • Calculation of Preliminary Net Advantages for Puerto Rico Complexes ... 125
Elements of Locational Advantage: the Full Programs 126
Elements of Locational Advantage: the Short Programs 132
Preliminary Evaluation of Programs 138
Effects of Alternative Wage-Rate Differentials 140

7 • Differential-Profitability Corrections: General Remarks ... 141
The Basic Types of Regional Nonidentities:
Their Interrelations 145
Differences in Process Reaction or Raw Material for a
Given Commodity or Intermediate: Associated
Product Differences 147
   Oil-Refinery Activities 147
   Petrochemical Activities 148
   Fiber Activities 150
Scale Differences 151
   Refinery Activities 151
   Petrochemical Activities 152
   Fiber Activities 153

CONTENTS   xi

CHAPTER   PAGE

    Technique for Calculating Total Differential-
    Profitability Corrections   153

8 • Differential-Profitability Analysis: Nylon **155**

    Nylon Salt Production Costs: Mainland Operations
    Identical with Puerto Rico Programs   156
        Plant-Investment Costs   156
        Raw-Material Cost Rates   156
        Use of Tables in Appendix C to Determine
        Raw-Material Rates   159
        Utility, Direct Labor, and Capital and Indirect Costs   160
    Nylon Salt Production Costs: Mainland Operations
    Achieving Maximum Scale Economies   161
        Choice of Output Scales for Specific Activities   161
        Estimated Production Cost, by Process   162
    Differential-Profitability Corrections: Maximum
    Mainland Economies of Scale   164
        Correction for Nylon Salt   164
        Corrections for Fertilizer Commodities   164
        Total Combined Correction   165
        Evaluation of Results   165
    Nylon Salt Production Costs: Mainland Operations
    Achieving Moderate Scale Economies   167
        Choice of Output Scales for Specific Activities   169
        Estimated Production Cost, by Process   170
    Differential-Profitability Corrections: Moderate
    Mainland Scale Economies   170
        Evaluation of Results   171
    Nylon Salt Production Costs: Mainland Operations
    Achieving Minimum Scale Economies   171
        Activity–Output Scales: Estimated Production Cost,
        by Process   173
    Differential-Profitability Corrections and Evaluation of
    Results: Minimum Mainland Scale Economies   175
    Summary   176

9 • Differential-Profitability Analysis:
Orlon, Dynel, and Dacron **178**

    Orlon   178
        Acrylonitrile Production Costs: Mainland Operations
        Identical with Puerto Rico Programs   178
        Choice of Output Scales for Specific Activities   180
        Differential-Profitability Corrections   181
        Evaluation of Results   183

xii CONTENTS

CHAPTER PAGE
>   Dynel   186
>   Acrylonitrile and Vinyl Chloride Production Costs:
>   Mainland Operation, by Scale-Economy Assumption   187
>   Differential-Profitability Corrections   189
>   Evaluation of Results   192
>   Dacron   192
>   Ethylene Glycol and Dimethyl Terephthalate
>   Production Costs: Mainland Operations, by
>   Scale-Economy Assumption   192
>   Evaluation of Results   195

10 · Modifications of the Analysis and Overall Conclusions **199**
>   Full Programs   199
>   Short Programs: Differential-Profitability Analysis   201
>   Adjusted Results: Short Programs   202
>   Continuous-Filament Fiber Programs   203
>   Effects of Different Assumptions on Wage-Rate
>   Differentials   206
>   Appraisal of Possible Methods for Reducing Scale
>   Disadvantages   208
>   Importation of One or More Chemical Intermediates   208
>   Production of Chemical Intermediates for
>   External Markets   210
>   Overall Conclusions   211

APPENDIX

A · Notes on Refinery and Ethylene-Separation Activities **217**

B · Specific Dynel, Dacron, Nylon Subprograms and Specific Orlon, Dynel, Dacron, Nylon Production Programs **232**

C · Estimated Production Costs: Chemical Intermediate Commodities **262**

Index **283**

List of Tables

Table		Page
1 •	Hypothetical Interregional Transactions Table, 19__, Cents Worth of Inputs per Dollar of Output	16–17
2 •	Input Requirements (Hypothetical), by Round, for $1 Million Output of Heavy Manufacturing in Region 1	20–21
3 •	Annual Inputs and Outputs for Selected Oil Refinery, Petrochemical, and Synthetic Fiber Activities	40–49
4 •	Capital and Labor Factors and Unit Level Requirements	54–55
5 •	Estimates of Indirect Costs Based on Direct Labor Costs and Capital Investment	59
6 •	Selected Indirect Costs: Expressed in Terms of Direct Labor, Imported and Local	60
7 •	Puerto Rico Fertilizer Imports, 1952, and Corresponding Quantities of Ammonia Equivalent in Nitrogen Content	65
8 •	Minimum Scales for Activities: in Multiples of Unit Level Outputs	68
9 •	Orlon Subprograms	78
10 •	Refinery Gas from Refinery Prototype #4 (Unit Level Operation): Hydrocarbon Composition and Alternative Uses	81
11 •	Total Requirements and Yields of Selected Industrial Complexes	90–91
12 •	Total Labor Requirements of Selected Industrial Complexes (100,000 MHR/YR)	94
13 •	Price (Cost and Revenue) Differences: Puerto Rico Less Mainland	98–99

LIST OF TABLES

Table		Page
14 •	Liquid Chemicals: Estimated Overseas Transport Cost between Puerto Rico and U.S. Mainland	111
15 •	Preliminary Locational Advantages of Selected Full Complexes in Puerto Rico	127
16 •	Preliminary Locational Advantages of Selected Short Programs in Puerto Rico	135
17 •	Preliminary Locational Advantages of Selected Full and Short Programs under Alternative Sets of Labor Price Differences (in $M/YR)	141
18 •	Production Cost of Nylon Salt, $/100 LB, Programs Identical with Puerto Rico	157
19 •	Activity Scales for Puerto Rico Nylon Programs and for Comparable Mainland Programs, by Scale Economies Assumption (in MM LB/YR)	158
20 •	Production Cost of Nylon Salt, $/100 LB: Maximum Scale Economies	162–163
21 •	Summary of Advantages and Disadvantages (in $/YR): Individual Puerto Rico Nylon Programs, by Mainland Scale Economies Assumption	166–169
22 •	Production Cost of Nylon Salt, $/100 LB, Moderate Scale Economies	172–173
23 •	Production Cost of Nylon Salt, $/100 LB, Minimum Scale Economies	174–175
24 •	Production Cost of Acrylonitrile, $/100 LB, Orlon Programs Identical to Puerto Rico	179
25 •	Activity Scales (in MM LB/YR) for Puerto Rico Orlon Programs and for Comparable Mainland Programs, by Scale Economies Assumption	180–181
26 •	Production Cost of Acrylonitrile (Orlon Programs), $/100 LB, by Scale Economies Assumption	182
27 •	Summary of Advantages and Disadvantages of Individual Puerto Rico Orlon Programs (in $/YR): by Mainland Scale Economies Assumption	184–185
28 •	Production Cost of Acrylonitrile, $/100 LB, Dynel Programs Identical to Puerto Rico	186
29 •	Production Cost of Vinyl Chloride, $/100 LB, Dynel Programs Identical to Puerto Rico	187

LIST OF TABLES

Table	Page
30 • Activity Scales for Puerto Rico Dynel Programs and for Comparable Mainland Programs, by Scale Economies Assumption (in MM LB/YR)	188–189
31 • Production Cost of Acrylonitrile (Dynel Programs), $/100 LB, by Scale Economies Assumption	190
32 • Production Cost of Vinyl Chloride (Dynel Programs), $/100 LB, by Scale Economies Assumption	191
33 • Summary of Advantages and Disadvantages of Individual Puerto Rico Dynel Programs (in $/YR): by Mainland Scale Economies Assumption	193
34 • Activity Scales for Puerto Rico Dacron Programs and for Comparable Mainland Programs: by Scale Economy Assumption (in MM LB/YR)	194–195
35 • Production Cost of Ethylene Glycol (Dacron Programs), $/100 LB, by Scale Economies Assumption	196
36 • Production Cost of Dimethyl Terephthalate (Dacron Programs), $/100 LB, by Scale Economies Assumption	197
37 • Summary of Advantages and Disadvantages of Individual Puerto Rico Dacron Programs (in $/YR): by Mainland Scale Economies Assumption	198
38 • Full and Reduced Puerto Rico Programs with Net Overall Advantages	200
39 • Scale and Process Disadvantages, Short Programs B and D (in $/YR)	202
40 • Overall Advantages of Selected Puerto Rico Short Programs (in $M/YR): by Mainland Scale Economies Assumption	204
41 • Overall Advantages of Selected Continuous Filament Programs (in $M/YR): by Mainland Scale Economies Assumption	205
42 • Overall Advantages of Selected Staple Fiber Programs (in $M/YR), by Mainland Scale Economies Assumption and under Each of Three Assumptions of Wage Rate Differentials	207
A–1 • Basic Operational Schemes of Refinery Prototypes	218
A–2 • Inputs and Outputs of Refinery Operations	220
B–1 • Dynel Subprograms	233–234
B–2 • Dacron Subprograms	235
B–3 • Nylon Subprograms	236

LIST OF TABLES

Table		Page
B- 4 to B- 8	Production Programs: Orlon	237–241
B- 9 to B-14	Production Programs: Dynel	242–247
B-15 to B-18	Production Programs: Dacron	248–251
B-19 to B-28	Production Programs: Nylon	252–261
C- 1	Production Cost of Ammonia, $/100 LB	266
C- 2	Production Cost of Nitric Acid, $/100 LB	267
C- 3	Production Cost of Ammonium Nitrate, $/100 LB	268
C- 4	Production Cost of Urea, $/100 LB	269
C- 5	Production Cost of Acetylene, $/100 LB	270
C- 6	Production Cost of Aromatics, $/BBL	271
C- 7	Production Cost of Butadiene, $/100 LB	272
C- 8	Production Cost of Calcium Carbide, $/TON	272
C- 9	Production Cost of Chlorine Gas, $/100 LB	273
C-10	Production Cost of Ethylene, $/100 LB	274
C-11	Production Cost of Ethylene Dichloride, $/100 LB	275
C-12	Production Cost of Ethylene Oxide, $/100 LB	276
C-13	Production Cost of Hydrogen Chloride from Chlorine and Hydrogen, $/100 LB 100% HCl	277
C-14	Production Cost of Hydrogen Cyanide, $/100 LB	278
C-15	Production Cost of Methanol, $/100 LB	279
C-16	Separation Cost of Paraxylene, $/GAL	280
C-17	Production Cost of Quicklime, $/100 LB	281
C-18	Production Cost of Sodium Cyanide, $/100 LB	281
C-19	Production Cost of Sulfuric Acid, $/100 LB	282

List of Figures

Figure	Page
1 • Flow sheet of principal petrochemical raw materials, intermediates, and end products	30–31
2 • Dacron A program	87
A-1 • Gas flows in the refineries	223
A-2 • Formalized gas flows, ethylene separation	226

List of Flowsheets

Flowsheet	Page
1 • Orlon	72
2 • Dynel	73
3 • Dacron	74
4A • Nylon	75
4B • Nylon	76
4C • Nylon	77

Summary

1. This study is addressed to the problem of (*a*) identifying some specific combinations of industrial activities for which Puerto Rico is likely to be an economically favorable location, and (*b*) estimating in dollars and cents terms the magnitude of the locational advantage of such combinations in Puerto Rico.

2. The specific combinations examined relate to complexes composed of oil refinery, petrochemical, fertilizer, and synthetic fiber operations. Numerous analytical techniques have been devised to deal with various aspects or problems of regional development, e.g., interregional input—output models, industry-by-industry comparative-cost studies, and extensive computation of diverse coefficients. None of these already existing methods is entirely satisfactory as a means of attacking the specific problem at hand. Accordingly, in this study a different analytical approach is developed. This approach combines an evaluation of interindustry relationships, along input-output lines, with an analysis of the factors that lead to regional differences in costs and revenues, along industry comparative-cost lines.

3. The first stage of the analysis is based on the cost comparison of a number of possible industrial complexes (combinations of industrial activities) in Puerto Rico with identical complexes on the mainland. Each complex involves the refining of crude oil and the production of refined petroleum products, petrochemical intermediates (including fertilizer components), and synthetic fiber materials. For each complex, total inputs by type and total outputs by type are calculated. For every basic input (e.g., crude oil, steam, fuel, labor) a cost differential between the Puerto Rico and the mainland locations is calculated. For every final output (e.g., gasoline, fuel oil, Dacron fiber) a revenue differential is obtained.

Algebraic summation of all the cost and revenue differentials of a given industrial complex yields the overall advantage or disadvantage of a Puerto Rico location over the mainland for that complex.

4. Given the provisional assumption that any complex in Puerto Rico would be operated in identical fashion on the mainland, each full complex considered emerges with a positive locational advantage for Puerto Rico. The programs calling for the production of Dacron result in generally greater advantage terms for the island than those including nylon, Orlon, or Dynel. The program Dacron A, consisting among other activities of a 30,000 bbl per day oil refinery, a 157 MM lb/yr fertilizer operation, and a 36.5 MM lb/yr synthetic fiber plant, is the most favorable one for Puerto Rico. It operates in Puerto Rico at a saving of $1.471 MM/yr.

5. The preliminary locational analysis makes clear that the petrochemical and to a lesser degree the refinery activities involve significant disadvantages for Puerto Rico (on skilled labor and transport account). On the other hand, Puerto Rico possesses pronounced advantages in the fertilizer and fiber activities (because of inexpensive textile labor and a substantial domestic demand for fertilizer). Accordingly, six alternatives to the full-length complexes are evaluated. These alternatives consist of refinery, fertilizer, or fiber activities alone, or in various combinations. Of these "short programs," the one combining fiber and fertilizer activities (Short Program D) is most favorable for Puerto Rico in terms of the provisional comparison with identical mainland operations. The preliminary net locational advantage of such a program in Puerto Rico is $2.287 MM/yr.

6. The detailed calculations of the study pertain to complexes producing staple fiber. If continuous-filament fiber production at the same scale is undertaken, significantly greater amounts of textile labor are required. On this account, Puerto Rico's advantage for every fiber-producing program is increased by $2.704 MM/yr. Thus, with continuous-filament production, Dacron A's preliminary locational advantage is $4.175 MM/yr; that of Short Program D, $4.991 MM/yr.

7. The second major stage of the analysis discards the unrealistic assumption that a mainland complex producing refinery, petrochemical, and synthetic fiber products would be identical with a complex producing the same products in Puerto Rico. It is recognized that in actuality there are likely to be important nonidentities between mainland and Puerto Rico operations. One is associated with differences in productive activities resulting from the use of

SUMMARY

different raw materials (e.g., oil and natural gas) or process reactions (e.g., oxidation and chlorhydrination of ethylene). Another is associated with differences in the output scales of the various intermediate or final commodities. The analytical method used to account for nonidentities is to compare the production costs of mainland operations, identical with the complexes set up for Puerto Rico, with the production costs of the same final commodities from mainland operations using the most efficient processes operated at the most efficient achievable scales. For each set of final commodities, the difference between these two mainland production costs represents the amount by which the preliminary net locational advantage of Puerto Rico for producing the corresponding commodities must be reduced. The corrections are calculated for three separate sets of assumptions as to the output scales possible on the mainland. Under the assumption that mainland operations can achieve "maximum" economies of scale, none of the staple fiber full programs shows a final net locational advantage for Puerto Rico.

However, in this situation the major scale disadvantages of Puerto Rico's petrochemical operations suggest the possibility of certain "reduced" full programs. These programs would be derived from full programs by scheduling one or more chemical intermediate commodities for import rather than production within the program. For example, the reduced Dacron A program is derived from the full Dacron A program by excluding from the latter the production of dimethyl terephthalate and scheduling this commodity for import. Such a change, given conditions of maximum mainland scale economies, eliminates a scale disadvantage of $908,000/yr. Thus, reduced Dacron A achieves a net locational advantage in Puerto Rico of approximately $73,000/yr.

If continuous-filament fiber is produced, several full programs achieve net locational advantage in Puerto Rico even under conditions of maximum mainland scale economies. Orlon J has an advantage of $2.135 MM/yr, and Dacron A has an advantage of $1.914 MM/yr. (Reduced Dacron A would have an advantage of $2.777 MM/yr.)

When mainland operations are assumed to achieve only "moderate" or "minimum" scale economies, several full programs producing staple fiber retain positive locational advantage for Puerto Rico. Of the ones that do, nylon, Dacron, and Orlon programs appear almost equally favorable. Under conditions of moderate mainland scale economies, Nylon G achieves an annual net overall advantage of $592,000; Orlon J, $468,000; and Dacron A, $311,000. If continuous-filament fiber is produced, each of these annual advantages is increased by $2.704 MM.

8. Differential-profitability corrections are also calculated for the short programs. Under conditions of maximum scale economies, Short Program A (fiber production alone) achieves the greatest advantage in Puerto Rico. The amount is $1.563 MM/yr for staple fiber, and $4.267 MM/yr for continuous filament. Under conditions of moderate mainland scale economies, Short Program D (fiber and fertilizer) has the greatest advantage. It amounts to $1.785 MM/yr for staple fiber and $4.489 MM/yr for continuous filament.

9. To provide a broader range of results for comparison, the effects of several variations in the assumptions affecting input and output price differences are examined. One variation completely eliminates the chemical–petroleum wage-rate disadvantage of Puerto Rico. Another cuts the chemical–petroleum wage-rate disadvantage of Puerto Rico in half and at the same time cuts in half the textile wage-rate advantage of Puerto Rico. Still another takes cognizance of possible discount on Venezuelan crude oil for a Puerto Rico producer. And so forth. In every case, at least some of the programs retain a positive advantage for Puerto Rico.

10. In summary, several sets of Puerto Rico–mainland price differences and differential-profitability hypotheses are considered in this study. For all but two of the situations postulated, there is at least one *full* or *reduced* program having positive advantage in Puerto Rico. For every situation postulated, there is at least one *short* program for which Puerto Rico has positive locational advantage.

Chapter 1

Possible Approaches to the Analysis of Puerto Rico as a Region[1]

INTRODUCTION

This book is concerned with Puerto Rico and the economic feasibility of developing in this region certain types of industrial activities—in particular, oil refining and petrochemical, fertilizer, and synthetic fiber production. It is also concerned with the process of economic growth and with the development of effective evaluation procedures for the identification of desirable channels of industrial expansion.

In attacking both these objectives, the authors were concerned with the selection of one or more techniques of regional analysis. It became clear after some investigation that several techniques of regional analysis are valid, each being pertinent for certain regional situations but not for others. This outcome was not unexpected. It is well known that for a long time economists, geographers, sociologists, political scientists, city and regional planners, and other social scientists have been concerned with the concept of "region." After much heated discussion and protracted writing, they have generally come to subscribe to a procedure that considers the region as a meaningful areal unit varying with the problem to

---

[1] This chapter is wholly methodological. The reader primarily interested in the calculations and conclusions of this study can proceed directly to Chapter 2.

be studied, the inclination of the investigator, and other features of a given situation. It is generally agreed that, until the "ultimate" is achieved in social science theory, analysts must be content with sets of regions—or hierarchies of sets of regions—which tend to differ from problem to problem. Given the existing state of social science knowledge, it is impossible to identify for all purposes a *best* set or *best* hierarchy of sets of regions.[2]

In like manner, the concept of regional structure has come to be relativistic. However a region is defined and whatever its extent, it includes certain broad components each of which pertains principally to one phase of organization of the region's society. Thus, the region has its economic sector, its politico-government sector, and its cultural-institutional sector, etc. Each of these broad components can be conceived as furnishing the framework or structure of the region. Furthermore, each major component consists of a number of units which can be combined or aggregated in varying degrees in accordance with different conceptual classifications. [For example, a region's economic sector might be described in terms of (1) the size range of individual firms, or (2) a classification of export and home industries, or (3) groups of firms or industries having strong production and marketing interconnections.] In practice, the nature of the particular problem to be investigated tends to determine the major component to be emphasized and the pattern of its disaggregation. Thus there can be many structural frameworks which are equally valid for the same region.[3]

Since both the concept and structure of a region vary with the problem and the existing situation, the analytical approach can be expected similarly to vary. There are many approaches to the study of economic development, and in particular to the identification of types of industries that might profitably develop in an underdeveloped region. Such approaches encompass the economist's broad theories of growth; comparative cost (industry-by-industry) analysis; the extensive use of various coefficients and quotients such as the location quotient, the labor coefficient, and the coefficient of localization; the study of commodity flows and balances of payments; the interregional input-output technique; linear programming; and the gravity model.

---

[2] See Walter Isard, "Regional Science, the Concept of Region, and Regional Structure," *Papers and Proceedings of the Regional Science Association*, Vol. II, 1956, pp. 13-26; and references cited therein.

[3] *Ibid.*, pp. 22-25.

Although each of these analytical approaches was evaluated for use in the present study of economic and industrial opportunities in the underdeveloped region of Puerto Rico, none as such was adopted. The following sections briefly describe these approaches and indicate why they were not adopted.

THE BROAD ECONOMIC DEVELOPMENT APPROACH

First to be considered are the broad theories of growth current in the field of economics. Although there is among economists no generally accepted comprehensive theory of economic development and growth, such an approach would place heavy emphasis upon concepts and phenomena such as capital-output ratios, savings and capital accumulation, resource and factor availability and allocation, investment levels, entrepreneurial innovation and expectations, balance-of-payments problems, disguised unemployment, and factor proportions and productivity. Also included might be more or less descriptive accounts of a chronological development process in a particular region, touching upon such things as changes in reproduction rates, mortality rates, and life expectancy, changes in the degree of urbanization both of population and of industry, and changes in the proportions of employment or value of product attributable to manufacturing and to service industries.[4]

Although for many purposes the approach just described is conceptually valid, it was not deemed suitable for this Puerto Rico investigation. A finer (albeit narrower) type of analysis was required—one that could accomplish the objective of identifying specific industries whose development in Puerto Rico might be justified in dollars and cents figures.

COMPARATIVE-COST ANALYSIS FOR INDIVIDUAL INDUSTRIES

The objective stated in the above paragraph logically leads to a consideration of a second approach, namely, an industry-by-industry comparative-cost analysis. Such an analysis is typically based, for

[4] For example, see Norman S. Buchanan and Howard S. Ellis, *Approaches to Economic Development,* Twentieth Century Fund, New York, 1955; Harold F. Williamson and John A. Buttrick, *Economic Development: Principles and Patterns,* Prentice-Hall, 1954; and *Capital Formation and Economic Growth,* National Bureau of Economic Research, Princeton University Press, 1955.

each industry studied, on an established or anticipated pattern of markets and a given geographic distribution of raw materials and other productive factors used in the industry. Within this framework the analyst attempts to determine in what region or regions the industry could achieve the lowest total cost of producing and delivering its product to market. Presumably, after a series of such studies of individual industries for which Puerto Rico could be a possible production location, the investigator would be able to identify those industries for which Puerto Rico tends to have positive locational advantages.

However, assimilation of various literature relating to Puerto Rico—e.g., reports of the Economic Development Administration of Puerto Rico—and perusal of the literature on the location of specific industries—e.g., Airov, *The Location of the Synthetic Fiber Industry*[5]—clearly point up the partial character of such studies.[6] They generally fail to treat interindustry relations in a thoroughgoing fashion. It is quite possible that, on the basis of typical individual industry location studies with respect to a given region, negative or inconclusive results regarding each industry might obtain, but that, in contrast, positive conclusions would be reached if interindustry relations were systematically considered. Such relations often lead to diverse external economies—economies stemming from a larger and more diversified labor force; from larger requirements of fuel, power, transport, and various urban and social facilities; from the localization in one district of diverse activities which feed by-products into each other, etc. The need to evaluate these and other similar relations tends to suggest an interindustry approach—e.g., input-output—to the problem of identifying possible types of industry development in Puerto Rico.

---

[5] Joseph Airov, *The Location of the Synthetic Fiber Industry*, John Wiley & Sons, 1959.

[6] Other individual industry studies that illustrate the comparative cost approach are: (1) John V. Krutilla, *The Structure of Costs and Regional Advantage in Primary Aluminum Production,* unpublished Ph.D. dissertation, Harvard University, 1952; (2) John Robert Lindsay, *The Location of Oil Refining in the United States,* unpublished Ph.D. dissertation, Harvard University, 1954, summary published as "Regional Advantage in Oil Refining," *Papers and Proceedings of the Regional Science Association,* Vol. II, 1956, pp. 304–317; (3) Walter Isard and William M. Capron, "The Future Locational Pattern of Iron and Steel Production in the United States," *The Journal of Political Economy,* **57,** 118–133 (1949) and (4) John Cumberland, *The Locational Structure of the East Coast Steel Industry with Emphasis on the Feasibility of an Integrated New England Steel Mill,* unpublished Ph.D. dissertation, Harvard University, 1951.

THE LOCATION QUOTIENT

Before a discussion of any interindustry approach, it should be pointed out that the problem could also be attacked by the calculation on a systematic and comprehensive basis of various sets of coefficients and quotients. For example, the investigator could calculate a location quotient for each of many different industries or activities, whether in Puerto Rico or not. A location quotient by definition is the ratio of a region's percentage share of a particular activity to its percentage share of some basic aggregate, the aggregate in the present context relating to the whole of the United States including Puerto Rico. The choice of the base aggregate depends on the nature of the problem—in this case the base aggregate might be total manufacturing employment or perhaps total population. For each industry, Puerto Rico's share of employment could be compared with her share of total manufacturing employment (on the mainland and in Puerto Rico).

At first thought, it would appear that industries with quotients greater than unity are "export" industries and those with quotients less than unity are "import" industries. Thus it might seem desirable to encourage and develop further the industries with quotients considerably greater than unity on the basis that these industries are areas of "strength" within the local economy. Also, it is tempting to urge the development of industries with location quotients considerably less than unity on the grounds that Puerto Rico does not have her "fair share" of such industries. However, these unqualified inferences can be seriously misleading. Because an individual region's consumption patterns, production practices, and income levels can vary widely from those of other regions and from the national average, a high (low) location quotient does not necessarily mean an export (import) industry. Thus, Puerto Rico might have a high location quotient for industry A because of above-average local consumption of its product. However, if production of industry A were stimulated further, it might be found that the excess production could not be exported except at a loss. Moreover, even if the industry is a major export industry, it does not follow that further expansion of exports is desirable; rather contraction of exports might be the sound course, as is true of the textile industry of New England.

For reasons of a somewhat similar character, a low location quotient does not guarantee the economic feasibility of increased local production. A ready local market may not exist. More significant, it might be desirable to encourage still further importation in order that Puerto Rico's resources be most effectively used and contribute most to an increase in the level of local income through greater labor productivity.

In sum, it is by no means automatically true that a wise path to industrial development is to encourage increased local production of all industries with significant export or import balances.[7]

THE COEFFICIENT OF LOCALIZATION

The limited information furnished by the location quotients for Puerto Rico could be supplemented to a certain extent by the calculation of a coefficient of localization for each industry. This measure essentially compares the percentage distribution by region of employment in the given industry with the percentage distribution by corresponding region of a base magnitude, e.g., total (national) manufacturing employment. Typically, the computation of the coefficient consists of summing over all regions the plus (or minus) deviations of the given industry distribution from the base distribution and dividing by 100. The limits to the value of the coefficient are 0 and 1; the higher the value, the greater the concentration of the location of the industry.

Some reasons for high coefficients of localization are the presence of marked economies of scale and the need to exploit a rich deposit of a raw material which is highly localized. Some factors leading to low coefficients of localization are the use of significant quantities of ubiquitous raw materials and heavy transport costs on finished product. Although a classification of industries by their coefficients of localization might be a useful preliminary step, it is evident that additional analysis is required to identify particular industries economically suited for development in Puerto Rico. Specific locational factors must be identified and evaluated in terms of potential revenues and costs in order to provide a firm dollar-and-cents rationale for action.[8]

On similar grounds, the study of commodity flows and balances

---

[7] For further discussion of location quotients, see Walter Isard et al., *Methods of Regional Analysis,* to be published.

[8] For further discussion of the coefficient of localization, see P. Sargant Florence, *Investment, Location, and Size of Plant,* Cambridge University Press, Cambridge, 1948, Chap. IV; and Isard et al., *op. cit.*

POSSIBLE APPROACHES TO THE ANALYSIS 11

of payments was eschewed. Maps and empirical presentation of commodity flows, and of corresponding financial flows and other items covered by balance of payments statements can be extremely valuable for descriptive purposes. They can neatly depict basic dimensions of the Puerto Rico situation and how these dimensions have changed over time; but in themselves they are of limited use in an attempt to identify possibilities for industrial expansion.[9]

THE LABOR COEFFICIENT

A step in the direction of considering specific locational factors could be taken by calculating for each industry a labor coefficient. Following Weberian location theory, one may define the labor coefficient of an industry as the ratio of labor cost per ton of product to the total weight of the localized raw materials and product involved in the production of a ton of product.[10] *Other things being equal,* industries with high labor coefficients would tend to find location in Puerto Rico more profitable than industries with low labor coefficients. The higher an industry's labor coefficient, the more likely it is that the labor-cost savings of a location in Puerto Rico would exceed any additional transport costs incurred by production at a site more distant from the market or raw materials source(s).

However, the phrase "other things being equal" severely limits the use of a rank of industries by their labor coefficients as an indicator of the relative degree of their attraction to a specific cheap labor area, even if labor costs and transport costs are the only significant locational influences. The most drastic condition required is that, in Weberian terminology, the location figures of all the industries should be identical. Essentially this means that, for each industry, location in the cheap labor-cost area would in-

---

[9] For further discussion, see Isard et al., *op. cit.*
[10] Calculations are based on "ideal" weights for all raw materials and product. Ideal weights are actual weights adjusted so as to effect an equalization of transport rates on all materials and product. For example, if the transport rate on raw material *A* is half that on the product *C*, the ideal weight of an actual ton of *C* is twice that of an actual ton of *A*. In the use of the labor coefficient, transport costs typically are assumed to be proportional to distance. For further details, see Carl J. Friedrich, *Alfred Weber's Theory of the Location of Industries,* University of Chicago Press, 1929, Chap. IV; and Walter Isard, *Location and Space-Economy,* John Wiley & Sons, 1956, Chap. VI.

volve the same net change in the effective distance over which raw materials and product are moved. Actually, such change commonly tends to vary substantially among industries. Two reasons for this are (1) the major markets and the best sources of raw materials tend to vary geographically from industry to industry, thereby leading to different minimum transport cost sites for different industries, and (2) the location, quantity, and accessibility of substitute raw-materials sources nearer the area of cheap labor may vary greatly among industries. It is possible that the net change in distance could be so different for two industries with reference to a specific cheap labor site that one, with a relatively low labor coefficient, could be *attracted* to the site, while the other, with a relatively high labor coefficient, would maintain a location at another site.

It becomes evident that under ordinary circumstances, at least a comparison for each industry of the labor-cost saving per unit output and the additional transport cost per unit output is required to provide a reasonably accurate assessment of the relative attraction of the cheap labor site for different industries. Such a calculation requires more information than the computation of a labor coefficient. Specifically, it is necessary to know (1) the amount by which wage rates differ in the cheap labor region from the rates used in the labor coefficient calculations, (2) the transport rate, and (3) for each industry the net effective change in distance involved in locating away from the original minimum transport-cost site.

In short, after the investigator properly qualifies the labor coefficient and supplements it with additional relevant materials, he has in effect performed a rather extensive comparative-cost computation. All the comments relating to the comparative-cost approach then become relevant here, too; the need to undertake an interindustry approach is not avoided.

THE INTERREGIONAL INPUT–OUTPUT APPROACH

When the present study was initiated, the input–output framework was the most prominent interindustry approach, in terms of both accomplishment and of recognition. When an input–output type of approach is considered with respect to the economic development of Puerto Rico, it becomes clear that such an approach should be interregional. The opportunities for Puerto Rico's development are, at least in the first instance, tied to mainland markets. Thus a fruitful attack would attempt to project growth of markets in the several mainland regions. Once the growth of regional markets is roughly identified, direct and indirect implications for industrial expansion in Puerto Rico can be approached via an interregional input-output model used in conjunction with comparative cost and other types of study.

If an interregional input-output scheme is to be adopted, an appropriate set of regions must be selected. Because of the small and compact nature of Puerto Rico, it is reasonable to consider the island as one major region. This view is also consistent with likely improvements in transportation and communications systems which will tend to transform the island into a "greater" metropolitan region with its focus at San Juan.

Next, the external relations of the island must be considered. Currently its major ties are with the metropolitan construct embracing the Greater New York-Philadelphia-Baltimore urban-industrial region. A lesser economic connection is with the Gulf Coast. Recognizing the difficulties of establishing new ties, and that institutional resistances and entrepreneurial inertia are among several forces tending to keep growth of economic activities within the framework of existing transportation and communication channels, one is inclined to anticipate that these two regions of the United States will continue to dominate the external relations of Puerto Rico.

There is a second ground for such belief. From a transport-cost standpoint, Puerto Rico is closer to both the Gulf Coast and the New York-Philadelphia-Baltimore urban-industrial region than to any other region of the United States. Even though in terms of physical distance the South Atlantic region is nearest the island, at best the likelihood is small that a sufficient volume of commodity movement will be generated between the South Atlantic region and Puerto Rico to realize the economies of scale, both in handling costs and in use of transport facilities, that are achieved in the Gulf Coast and New York-Philadelphia-Baltimore trade. This signifies that from an economic standpoint the South Atlantic region is considerably more distant.

Moreover, Puerto Rico, as a growing economy, is likely to find that the sale of additional output through displacing existing suppliers in a well-established market is more difficult than through capitalizing on new market demand. Because the Gulf Coast and the New York-Philadelphia-Baltimore areas will be among the most rapidly expanding regions of the United States, it does not appear unreasonable to expect that Puerto Rican businessmen will concentrate for the most part on these two regions for new sales outlets.

Hence, if the problem is to project the interrelations between Puerto Rico and the mainland and if research resources limit consideration to only three regions, the Gulf Coast and the New York-Philadelphia-Baltimore urban-industrial area may be considered the most significant external regions for the analysis.[11]

Once a set of regions is chosen, it becomes necessary to select for each region the most appropriate group of industries (industrial classification). For Puerto Rico, specific industries such as sugar cane growing, sugar processing, needlework, synthetic fiber manufacture, and fertilizer production would tend to be explicitly considered. Activities such as steel making, aluminum manufacture, and ore mining, would not appear individually because of their relative unimportance, currently and prospectively.[12] For the Gulf Coast area, petroleum refining and petrochemical production might appear explicitly in an industrial classification, whereas textile manufacture could be consolidated with numerous other categories of minor significance. With respect to the New York-Philadelphia-Baltimore region, activities such as steel manufacture, steel fabrication, and textile production would appear explicitly.

Because the present investigation did not undertake an interregional input-output study, it is not possible at this juncture either to present the most appropriate set of industrial sectors for each region or to identify the transactions corresponding to flows among these industrial sectors. However, it is possible to design a hypothetical case to illustrate how a three-region Puerto Rico model might work.[13] Accordingly, Table 1 has been constructed.

---

[11] If a large number of regions could be treated, the Pacific Coast and the "Rest of the United States" might qualify as meaningful areas.

[12] For information pertaining to an appropriate industrial classification for Puerto Rico, and in general to input-output analysis for Puerto Rico, see A. Gosfield, "Input-Output Analysis of the Puerto Rican Economy," in *Input-Output Analysis: an Appraisal,* Studies in Income and Wealth, Vol. 18, Princeton University Press, 1955, pp. 321-362; and *How to Select Dynamic Industrial Projects,* U. S. International Cooperation Administration, Office of Industrial Resources, Washington, 1956, pp. A31-A51.

[13] The presentation of this hypothetical case draws heavily upon and to a great extent is taken verbatim from Walter Isard and R. Kavesh, "Economic Structural In-

POSSIBLE APPROACHES TO THE ANALYSIS 15

Table 1 refers to three hypothetical regional economies, each broken down into nine sectors. To avoid cumbersome detail and to simplify the discussion, the table is so designed that for each region all export activities are consolidated into one sector—heavy manufacturing for region I, light manufacturing for region II, and agriculture and extractive activity for region III. The next eight sectors are identical for each region: namely, power and communications, transportation, trade, insurance and rental activities, business and personal services, educational and other basic services, construction, and households.[14] Each of these eight activities is for the moment assumed to be local in nature. None of their output is shipped outside the region in which it is produced. Thus, by definition, it is through export activities alone that the simplified economies of the several regions are interrelated.

Classification of outputs represents only one phase of the problem. Another phase concerns input structures; more specifically, the inputs of each of several factors—raw materials, power, transportation, labor, equipment, and other services—required to produce a unit of output. In actuality, much of the output of many industries such as basic steel is absorbed by other industries as inputs rather than by households. Therefore, in order to understand the economic base of regions and to anticipate changes within them, it is necessary to know the interregional input structures of industries. This requires a table of interregional interindustrial relations for a base-year period, on the order of Table 1.[15]

---

terrelations of Metropolitan Regions," *American Journal of Sociology,* LX, 152-162 (1954), an article that represented a preliminary report on certain thinking in connection with this project. Additional technical details on the workings of the interregional model to be presented appear in Robert A. Kavesh, *Interdependence and the Metropolitan Region,* unpublished Ph.D. dissertation, Harvard University, 1953.

[14] The output of households roughly corresponds to the value of the services of labor and of capital and land owned by them.

[15] Most of the coefficients in Table 1 are based upon a consolidation of the 50 × 50 interindustry flow matrix developed by the Bureau of Labor Statistics. See W. D. Evans and M. Hoffenberg, "The Interindustry Relations Study for 1947," *Review of Economics and Statistics,* XXXIV, 97-142 (1952). In reducing the B.L.S. 50-industry classification to our three export and eight local industrial categories we crudely defined:

1. *Heavy manufacturing* as the aggregate of iron and steel; plumbing and heating supplies; fabricated structural metal products; other fabricated metal products; agricultural, mining, and construction machinery; metalworking machinery; other machin-

INDUSTRIAL COMPLEX ANALYSIS

TABLE 1
HYPOTHETICAL INTERREGIONAL TRANSACTIONS TABLE, 19——,

				Industry Purchasing					
		Region I							
Industry Producing	Heavy Manufacturing (1)	Power and Communication (2)	Transportation (3)	Trade (4)	Insurance and Rental (5)	Business and Personal Services (6)	Educational and Other Services (7)	Construction (8)	Households (9)
Region I:									
1. Heavy Manufacturing	33	1	3	1	....	9	1	18	3
2. Power and communication	1	11	3	2	8	4	2	....	1
3. Transportation	2	2	5	1	1	1	2	4	3
4. Trade	1	....	2	....	2	3	5	9	12
5. Insurance and rental activities	1	1	3	5	7	5	4	2	12
6. Business and personal services	1	1	2	7	1	4	2	3	3
7. Educational and other basic services	....	....	....	....	....	....	1	....	10
8. Construction	....	4	6	....	10	....	1	....	....
9. Households	34	58	58	63	53	46	50	40	1
Region II:									
10. Light Manufacturing	4	1	2	2	1	14	15	4	20
11. Power and communication	....	....	....	....	....	....	....	....	....
12. Transportation	....	....	....	....	....	....	....	....	....
13. Trade	....	....	....	....	....	....	....	....	....
14. Insurance and rental ac.	....	....	....	....	....	....	....	....	....
15. Business and personal services	....	....	....	....	....	....	....	....	....
16. Educational and other basic services	....	....	....	....	....	....	....	....	....
17. Construction	....	....	....	....	....	....	....	....	....
18. Households	....	....	....	....	....	....	....	....	....
Region III:									
19. Agriculture and extraction	6	5	4	1	2	....	4	18	6
20. Power and communication	....	....	....	....	....	....	....	....	....
21. Transportation	....	....	....	....	....	....	....	....	....
22. Trade	....	....	....	....	....	....	....	....	....
23. Insurance and rental ac.	....	....	....	....	....	....	....	....	....
24. Business and personal services	....	....	....	....	....	....	....	....	....
25. Educational and other basic services	....	....	....	....	....	....	....	....	....
26. Construction	....	....	....	....	....	....	....	....	....
27. Households	....	....	....	....	....	....	....	....	....

ery (except electric); motors and generators; radios; other electrical machinery; motor vehicles; other transportation equipment; professional and scientific equipment; miscellaneous manufacturing; and scrap and miscellaneous industries.

2. *Light manufacturing* as the aggregate of food and kindred products; tobacco manufactures; textile mill products; apparel; furniture and fixtures; paper and allied

TABLE 1
CENTS WORTH OF INPUTS PER DOLLAR OF OUTPUT

	Industry Purchasing — Region II									Industry Purchasing — Region III								
	Light Manufacturing	Power and Communication	Transportation	Trade	Insurance and Rental	Business and Personal Services	Educational and Other Services	Construction	Households	Agriculture and Extraction	Power and Communication	Transportation	Trade	Insurance and Rental	Business and Personal Services	Educational and Other Services	Construction	Households
	(10)	(11)	(12)	(13)	(14)	(15)	(16)	(17)	(18)	(19)	(20)	(21)	(22)	(23)	(24)	(25)	(26)	(27)
	2	1	3	1	....	9	1	18	3	1	1	3	1	....	9	1	18	3
	....	....	....	....	....	....	....	....	....	....	....	....	....	....	....	....	....	....
	....	....	....	....	....	....	....	....	....	....	....	....	....	....	....	....	....	....
	....	....	....	....	....	....	....	....	....	....	....	....	....	....	....	....	....	....
	....	....	....	....	....	....	....	....	....	....	....	....	....	....	....	....	....	....
	....	....	....	....	....	....	....	....	....	....	....	....	....	....	....	....	....	....
	....	....	....	....	....	....	....	....	....	....	....	....	....	....	....	....	....	....
	28	1	2	2	1	14	15	4	20	6	1	2	2	1	14	15	4	20
	1	11	3	2	8	4	2	....	1	....	....	....	....	....	....	....	....	....
	2	2	5	1	1	1	2	4	3	....	....	....	....	....	....	....	....	....
	2	....	2	....	2	3	5	9	12	....	....	....	....	....	....	....	....	....
	1	1	3	5	7	5	4	2	12	....	....	....	....	....	....	....	....	....
	2	1	2	7	1	4	2	3	3	....	....	....	....	....	....	....	....	....
	....	....	....	....	....	....	1	....	10	....	....	....	....	....	....	....	....	....
	....	4	6	....	10	....	1	....	....	....	....	....	....	....	....	....	....	....
	25	58	58	63	53	46	50	40	1	....	....	....	....	....	....	....	....	....
	21	5	4	1	2	....	4	18	6	28	5	4	1	2	....	4	18	6
	....	....	....	....	....	....	....	....	....	1	11	3	2	8	4	2	....	1
	....	....	....	....	....	....	....	....	....	3	2	5	1	1	1	2	4	3
	....	....	....	....	....	....	....	....	....	2	....	2	....	2	3	5	9	12
	....	....	....	....	....	....	....	....	....	4	1	3	5	7	5	4	2	12
	....	....	....	....	....	....	....	....	....	1	1	2	7	1	4	2	3	3
	....	....	....	....	....	....	....	....	....	....	....	....	....	....	....	1	....	10
	....	....	....	....	....	....	....	....	....	....	4	6	....	10	....	1	....	....
	....	....	....	....	....	....	....	....	....	40	58	58	63	53	46	50	40	1

products; printing and publishing; chemicals; rubber products; and leather and leather products.

3. *Agriculture and extraction* as the aggregate of agriculture and fisheries; lumber and wood products; products of petroleum and coal; stone, clay, and glass products; and nonferrous metal.

In Table 1 any one column records the cents' worth of inputs from each industrial category in each region per dollar's worth of output of a given industrial category of a given region where both the given industrial category and the given region are specified by the column heading. For example, reading down column 1 furnishes information on the cents' worth of various inputs from the several regions per dollar output of heavy manufacturing in region I. Thirty-three cents' worth of heavy manufacturing in region I is fed back as an input into the same activity in the same region for every dollar's worth of its output (such as Pittsburgh steel, which is fed back to Pittsburgh steelworks). Two cents of transportation services of region I is absorbed per dollar's worth of heavy manufacturing of region I. In addition to inputs from other service sectors and the household sector of region I, the heavy manufacturers of region I require inputs from the light-manufacturing industry of region II and from agriculture and extractive activities of region III. These latter, of course, entail interregional flows.

Consider another column, the fifteenth, which refers to "Business and Personal Services" in region II. Per dollar of its output 9 cents' worth of heavy-manufacturing products from region I is required. None of the other sectors of region I furnishes inputs, because these other sectors are defined as local and hence export nothing. Since the business and personal services sector of region II does not consume any agricultural and extractive products, all its other inputs must come from region II, as is depicted in Table 1.

Service activities were expressed in a less aggregative form in order to present some detail on the internal structural processes of metropolitan regions associated with these activities. The category, "Education and Other Basic Services," consists of the services of medical, educational and nonprofit institutions; amusement; and eating and drinking places.

Certain activities are omitted from the analysis because their levels of output are not structurally related to the interindustrial matrix of coefficients. These are inventory change, foreign trade, government, capital formation, and unallocated. Households, generally included with this group, were introduced into the structural matrix in order to catch the local multiplier effect of new basic industry upon a region via the additional income generated.

The actual derivation of a coefficient involves the division of the total value of inputs from a given sector into a second sector by the output of the second sector. That is, if in 1947 the amount of chemicals used in steel production was $99 MM and the output of steel was $12.3 MMM, the input coefficient representing the cents' worth of chemicals per dollar of steel would be 0.8049.

The data are rounded to the nearest whole figure. Inputs of less than ½¢/$ output are not recorded.

POSSIBLE APPROACHES TO THE ANALYSIS 19

Aside from their obvious descriptive value, of what significance are data such as those of Table 1? In general, input structures are not haphazard; rather they reflect to a large extent stable and meaningful relations. If the output of an efficient aluminum works is doubled, it is reasonable to expect that approximately twice as much power, alumina, carbon electrodes, and other inputs will be required. In short, subject to certain serious qualifications to be discussed later, the input of any service or good into a particular activity may be said to vary approximately, within certain limits, in direct proportion to the output of that particular activity.

To illustrate the usefulness of input structure information, suppose that a resource development program calls for an increase of one million dollars in the output of heavy manufacturing in region I. How will this affect the output of each activity in each region?

In column 1 of Table 1 are listed coefficients which indicate the cents' worth of various inputs required per dollar output of heavy manufacturing. Multiplying these coefficients by one million gives the direct inputs required to produce one million dollars' worth of heavy manufactures. These are called the first-round input requirements and are listed in column 1 of Table 2.

But to produce the first-round requirement of $330,000 of heavy manufacturing (item 1 in column 1, Table 2) likewise requires a whole series of inputs. These can be obtained by multiplying column 1 of Table 1 by 330,000. And to produce the $20,000 of transportation (item 3, column 1, Table 2) requires inputs that can be obtained by multiplying column 3 of Table 1 by 20,000. Similarly, the inputs required to produce each of the other items listed in column 1 of Table 2 can be derived. It should be noted that the $340,000, which is listed in the ninth cell of column 1, Table 2, represents an increment of income received by the households in region I. This increment results in increases in effective demand for a series of products. On the artibrary assumption that two-thirds of this new income is spent, these increases in effective demand can be obtained by multiplying column 9, Table 1 (which shows how a dollar spent by households is typically distributed among various products), by 226,667.

Adding together all these inputs (including the new effective demands of households) necessary for the production of the first round of requirements yields the second round of requirements, which is recorded in column 2 of Table 2. In turn, the production of the second round of requirements necessitates a third round. This is computed in the same manner as was the second round.

TABLE 2
INPUT REQUIREMENTS (HYPOTHETICAL), BY ROUND,

Industry Grouping	First-Round Input Requirements (1)	Second-Round Input Requirements (2)	Third-Round Input Requirements (3)
Region I:			
1. Heavy Manufacturing	$330,000	$118,810	$ 47,793
2. Power and Communication	10,000	8,670	7,763
3. Transportation	20,000	14,910	7,417
4. Trade	10,000	31,440	15,687
5. Insurance and rental activities	10,000	32,940	18,965
6. Business and personal services	10,000	11,810	8,159
7. Educational and other basic services		22,700	10,077
8. Construction		2,600	4,759
9. Households	340,000	148,070	110,102
Region II:			
10. Light manufacturing	40,000	75,600	60,601
11. Power and communication		400	971
12. Transportation		800	1,781
13. Trade		800	2,364
14. Insurance and rental activities		400	1,696
15. Business and personal activities		800	1,825
16. Educational and other basic services			670
17. Construction			104
18. Households		10,000	20,747
Region III:			
19. Agriculture and extraction	60,000	60,220	50,741
20. Power and communication		600	1,122
21. Transportation		1,800	2,430
22. Trade		1,200	3,226
23. Insurance and rental activities		2,400	4,646
24. Business and personal services		600	1,256
25. Educational and other basic services			1,600
26. Construction			372
27. Households		24,000	27,936
Total	$830,000	$571,570	$414,810

Furnishing a third round requires a fourth; a fourth round, a fifth; etc. Each of these rounds is recorded in Table 2. It should be noted that the totals of the rounds converge.[16] After a certain point is reached, it becomes feasible to stop the round-by-round computation and to extrapolate the remaining requirements. However, we have not carried through any extrapolation; as a refinement it implies a degree of accuracy and stability in the data which, as we shall see in the following section, does not exist in fact.

Thus, we have developed a round-by-round description of how an impulse acting upon one sector of a region is transmitted to every sector in the same region and every other region. To derive the total effect, it is merely necessary to sum the rounds horizontally. The totals are recorded in column 8 of Table 2. These

---

[16] The convergence of rounds results from the assumption that only two-thirds of the income received by households in any given round is expenditure in the succeeding round, and from the omission of the nonstructurally related sectors of inventory change, foreign trade, government, capital formation, and unallocated, as noted in the preceding footnote.

TABLE 2
FOR $1 MILLION OUTPUT OF HEAVY MANUFACTURING IN REGION I

Fourth-Round Input-Requirements (4)	Fifth-Round Input Requirements (5)	Sixth-Round Input Requirements (6)	Seventh-Round Input Requirements (7)	Sum of Rounds (8)
$ 23,417	$ 13,407	$ 8,559	$ 5,884	$550,870
4,614	2,858	1,667	994	36,566
4,508	2,516	1,475	871	51,697
11,021	6,042	3,573	2,060	79,823
12,612	7,135	4,155	2,430	88,237
4,860	2,906	1,664	983	40,382
7,463	3,945	2,359	1,344	47,888
2,731	1,789	1,031	622	13,532
57,920	34,886	19,773	10,805	721,556
47,894	34,849	25,264	18,115	302,323
1,182	1,190	1,056	856	5,655
1,821	1,601	1,309	1,016	8,328
3,044	2,858	2,470	1,963	13,499
2,689	2,706	2,490	1,972	11,953
1,954	1,772	1,479	1,159	8,989
1,387	1,394	1,275	1,033	5,759
325	446	455	391	1,721
20,643	18,918	15,744	12,381	98,433
39,365	29,244	21,250	15,387	276,207
1,402	1,386	1,229	1,019	6,758
2,360	2,085	1,673	1,310	11,658
3,541	3,481	2,922	2,385	16,755
4,962	4,701	3,917	3,156	23,782
1,490	1,463	1,260	1,032	7,101
1,876	1,969	1,680	1,397	8,522
664	719	682	581	3,018
28,508	25,037	20,595	16,189	142,265
$284,253	$211,303	$151,006	$107,335	$2,583,277

totals of course, can be compared with other sets of totals which reflect impacts of other types of impulses.[17]

However, this simplified model must be qualified and fashioned somewhat more realistically.[18] It has already been noted that a much more relevant industrial classification should be adopted for

---

[17] For example, if instead of $1 MM of new heavy manufacturing, an equivalent amount of new agricultural and extractive output is required, the impact will be more localized and confined to the region of initial expansion (region III). For full details and other contrasts see Kavesh, *op. cit.,* Chap 3.

[18] Because of limitation of space, we shall discuss only briefly the several important points that are raised. Full discussion of these points is contained in W. Leontief, *The Structure of the American Economy, 1919-1939,* Oxford University Press, New York, 1951; W. Leontief et al., *Studies in the Structure of the American Economy,* Oxford University Press, New York, 1953; Walter Isard, "Interregional and Regional Input-Output Analysis: A Model of a Space-Economy," *Review of Economics and Statistics,* **XXXIII**, 318-328, (1951); Walter Isard, "Regional Commodity Balances and Interregional Commodity Flows," *American Economic Review,* **XLIII**, 167-80, (1953); Walter Isard, "Location Theory and Trade Theory; Short-Run Analysis," *Quarterly Journal of Economics,* **LXVIII**, 305-20, (1954); and *Input-Output Analysis: an Appraisal,* Studies in Income and Wealth, Vol. 18, Princeton University Press, 1955.

each region. Not one, but several or many of the industries of each region would be export industries, and thereby would interrelate the several regions. Yet, it must be recognized that however good the industrial classification for each region, shortcomings arise because in reality most industries produce several commodities, and computation resources limit the fineness of any industrial classification that can be employed.[19]

Apart from these shortcomings, there is the question of stability of input coefficients—the assumption that from round to round the cents' worth of any input per unit of a given output remains constant, or the equivalent, namely, that the amount of any input supplied an industry varies proportionally with the output of that industry. As the output of an industrial activity expands, new combinations of the various inputs and new technical processes may become economically feasible. These new combinations and processes would require different percentage increases in the various inputs into the production process; this would be inconsistent with the basic assumption. For many industries such changes might involve minor substitutions of one type of input for another and hence might not significantly bias the results. In other industries there may be major substitution effects. However, to the extent that these effects can be anticipated, they can be incorporated into the model by the appropriate alteration of coefficients in the relevant rounds.

The stability of input coefficients is particularly to be questioned in a region such as Puerto Rico which is still relatively underdeveloped. It is quite likely that, as plants take root in Puerto Rico, new techniques will be used, especially since incipient industrialization has a significant effect on the attitudes of the working force, which in turn is reflected in labor productivity.[20] As a result, it is necessary to secure for such new plants the set of inputs that prospective management may expect to be required for current operation and/or to approximate from social science research studies the effects of the introduction of new industry upon labor productivity and in turn upon the set of inputs and techniques utilized. Obviously, where no adequate information is available, it becomes necessary to rely heavily upon individual judgment.

[19] See M. Holzman, "Problems of Classification and Aggregation," in W. Leontief et al., *op. cit.*, Chap IX.

[20] See, for example, W. E. Moore, *Industrialization and Labor,* Cornell University Press, 1951.

POSSIBLE APPROACHES TO THE ANALYSIS 23

Associated with the above shortcoming are the restraints which limited resources impose. For example, as the demand for coal rises, veins of an inferior quality may need to be exploited. This in turn would lead to greater consumption of coal per unit of output of a coal-consuming industry. At the extreme, where there are fixed limits to a given resource (including human labor services), entirely new production techniques and/or locations may be dictated to realize increments of output.[21] Again, to the same extent that resource limitations and associated changes in production techniques can be anticipated, so can the coefficients for the several rounds be altered to incorporate into the analysis relevant information on these factors.

Still more critical a qualification stems from changes in consumption patterns incident to income changes.[22] Simple cents' worth of inputs per dollar of income, which are listed in columns 9, 18, and 27 of Table 1, are misleading. Consumers studies are required, in which households are broken down by occupation, ethnic grouping, family size, rural–urban location, and other key indicators, to reveal how expenditure patterns are related to changes in the level of income and associated occupational shifts. Once obtained, relevant information can be injected into the model to yield more valid results.

Again, from the standpoint of an underdeveloped region such as Puerto Rico, this limitation of the model is still more severe. Major nonlinearities in consumption habits are to be anticipated as

---

[21] The data presented in the above tables are expressed in dollars and cents. Yet they can be easily translated into physical units. For example, consider the labor problem in a given market area. It is possible to introduce new rows in Table 1, where each row corresponds to a particular type of labor (skilled, semiskilled, manual, etc.), the nature of the problem determining the particular breakdown of labor to be adopted. Reading down any column would denote the requirements of each type of labor (in terms of man-hours) to produce a unit of output corresponding to the industry and region listed at the head of the column. Thus, in studying the impact of any given resource development program, we can derive the additional requirements of various types of labor by regions; this in turn throws light not only on the short-run feasibility of a given resource development program but also on the likely long-run interregional labor migrations (given information on reproduction rates and other population characteristics). In similar fashion, a conversion of the table to physical terms could supply insights on the adequacy of actual power facilities, housing, and transportation networks of various regions.

[22] The socioeconomic data basic to Engel's law indicate this tendency. For discussion of this law see, among others, Carle C. Zimmerman, *Consumption and Standards of Living,* D. Van Nostrand Co., 1956, and S. J. Prais, "Non-linear Estimates of the Engel Curves," *Review of Economic Studies,* **XX,** 87–104, (1952–53).

income rises. Unfortunately, data are relatively sparse on how industrialization, increasing urbanization, and intensified contact with the mainland might influence cultural patterns of the island, to say nothing of how such institutional factors as entrepreneurial vigor and savings schedules might be modified.

Another major set of qualifications is linked to the resource limitations already noted. As long as there is vacant housing around an industrial district of a region, excess capacity in the transit and power systems, and available space for expansion, the calculated growth of the industrial centers can be effected. However, where vacant housing does not exist and where streets are congested and transit and power systems overloaded, additional capacity must be constructed to permit expansion in the various industries and service trades. Therefore, in addition to the inputs that are required to produce expanded outputs from existing and new facilities, a whole series of inputs is required to construct the new facilities.

Here, too, appropriate modification of the model can be made. For example, given a knowledge of the capacity of an existing housing complex (together with information on the doubling-up effect and other cultural adaptations to shortage known to be feasible, the nature of the demand for diverse types of housing, and the input structures of the several sectors of the housing industry), it is possible to allow for the phenomenon of housing expansion in the analytic framework. It should be borne in mind, however, that to the extent to which a particular resource in short supply is diverted, from producing output on current account, to building up plant equipment and other capacity to produce, then to a similar extent the expansion of the noncapacity-building activities are curtailed.[23] Furthermore, another set of nonlinearities is introduced when we consider the problem of plant scale in underdeveloped countries such as Puerto Rico. In many cases the justification for erecting a plant in a given industry is lacking because the potential market is inadequate for absorbing the output of a plant of a minimum technically feasible size. However, as development proceeds and effective demand mounts, a stage may be reached where demand does become adequate for a particular market-oriented operation, such as cement production. When

---

[23] For a treatment of the problem underlying these assumptions, see *Problems of Capital Formation,* Studies in Income and Wealth, Vol. 19, Princeton University Press, 1957, Part II.

POSSIBLE APPROACHES TO THE ANALYSIS

effective demand does reach such a level, it becomes necessary to alter the entire set of technical coefficients relating the input of the given commodity, say, cement, from any given region into each industrial activity of every region. This and similar alterations can be effected in round-by-round computations if, beforehand, information relating to such potential shifts is available.

Finally, with respect to potential industrial development in Puerto Rico based on mainland markets, it would be desirable to change certain coefficients in the light of results from comparative-cost studies. For example, Puerto Rico currently (1959) does not produce synthetic fibers. Relevant comparative-cost studies may indicate that such activity ought to be located in Puerto Rico to serve mainland markets. If so, appropriate adjustments of coefficients reflecting inputs of Puerto Rican synthetic fiber and Southern synthetic fiber, etc., per dollar's worth of Southern textiles and Northern textiles, etc., would have to be effected to reflect these cost conclusions.

Thus, in the above manner, the initial, highly simplified linear model—linear in the sense that each input varies in direct proportion to the output—can be molded into a less hypothetical, nonlinear model which does recognize important nonproportionalities in interactivity relations. As indicated, there would be many serious problems arising from, among others, the inadequacies of the data, the unpredictability of changes in behavioral patterns and culture, the uncertain direction and magnitude of technological development, and the need for a number of important comparative-cost studies whereby to effect the required changes in interregional input–output coefficients. In such a model there would be little feedback from Puerto Rico to the mainland. In essence the basic problem would be to make projections for the relevant regions of the mainland, to investigate what new demand in these regions might be met from Puerto Rico production, and to trace the multiplier effect on the Puerto Rico region.

SOME REASONS FOR THE CHOICE OF AN INDUSTRIAL-COMPLEX APPROACH

Along the lines of the model sketched above, it would have been possible to have constructed a Puerto Rico interregional input–output model. However, after carefully weighing the several virtues and limitations of such a model, the authors judged that, given the limited resources available for research, it would be more profitable to investigate an alternative interindustry approach. As indicated above, the identification of types of industrial development favorable for Puerto Rico is necessary for the construction of interregional input–output coefficients capable of meaningfully depicting Puerto Rico's future development. Thus it was decided to undertake a detailed study of a few strongly interrelated sectors, in which some of the breadth of input–output would be sacrificed for a penetrating and thoroughgoing cost analysis of the sectors selected. But, as already noted, the possibility exists that orthodox comparative-cost analysis of each sector individually might indicate negative or inconclusive results for Puerto Rico when analysis of these sectors in combination might yield positive results. Such a situation can arise because of economies of scale and integration, urbanization economies, by-product relations, and other positive agglomeration factors.[24] This consideration led to the industrial-complex analysis which is presented in detail in succeeding chapters. Needless to say, the analysis cannot express quantitatively all possible types of economies which may be associated with an industrial complex. We purport to cover only the more significant economies associated with complexes based on hydrocarbon raw materials.

To bring this chapter to a close, a few words are in order regarding the use of linear programming techniques and gravity models. At the time of initiation of this study, these two approaches were not sufficiently developed and proved to warrant their use; hence neither was seriously considered. Moreover, the linear programming approach, in many respects similar to the input–output approach, pertains especially to situations wherein resource limitations are pressing. The authors judged that such limitations would not particularly influence the growth of those industries for which Puerto Rico might possibly have advantage. This provided another basic reason for not considering the use of linear programming analysis, and for proceeding with the development of an industrial-complex approach.[25]

---

[24] Since the input–output approach cannot evaluate such factors, this was one of the important reasons for not pursuing it.

[25] However, see the subsequent work by Thomas Vietorisz, *Regional Programming Models and the Case Study of a Refinery–Petrochemical–Synthetic Fiber Industrial Complex for Puerto Rico,* unpublished Ph.D. dissertation, M.I.T., 1956.

Chapter 2

The Choice of Relevant Industrial Complexes

THE RESOURCE SETTING

As stated in Chapter 1, this study was undertaken in order to identify types of industrial activities that might profitably locate in Puerto Rico. In such a regional study, a typical first step involves a reconnaissance of resources available for use. In the case of Puerto Rico, one basic resource that is quickly identified is the large supply of moderately skilled, cheap labor. Highly skilled labor, on the other hand, does not exist in abundance and is obtained generally at a disadvantage relative to other regions. Nonhuman natural resources are for the most part meager. Industrial minerals including coal, as well as precious metals, are generally scarce and too low in quality to permit commercial exploitation. In spite of some development of hydroelectric power and the use of sugar cane waste as fuel for power generation, a large part of the island's power is based on imported fuel oil. Fish and forest resources are strictly limited and inadequate. Land is scarce relative to population, and only a small proportion of the soil is of high quality; acceptable crop yields generally require the use of significant amounts of fertilizer.[1]

A more penetrating examination of Puerto Rico's situation, in particular an examination that proceeds within an interregional framework, reveals at least a second and a third major resource of

---

[1] For a more detailed enumeration of Puerto Rico's resources see H. S. Perloff, *Puerto Rico's Economic Future,* University of Chicago Press, 1950, pp. 45–53.

the island. The second is the island's proximity to Venezuelan oil. In terms of geographic distance, Puerto Rico is 400 to 500 miles from rich Venezuelan oil deposits. However, since this distance is over water, and since water shipment is much less costly than other forms of shipment, Puerto Rico's economic distance from Venezuela is much less than 400 to 500 miles of overland separation.

A third important resource of Puerto Rico is its free (dutyless) entry and easy access to mainland markets of the United States. This resource affords Puerto Rico a major advantage relative to other cheap labor regions of the world, including some within the United States.

INDUSTRIAL ACTIVITIES SELECTED

Given these three chief resources, how can one combine them by means of a matrix of industrial activities to yield the maximum advantage for a Puerto Rico location?

Let us begin the analysis with Figure 1. This figure presents in flowsheet form some of the many commodities that stem from crude oil and natural gas, as well as some of the processes by which those commodities are produced.[2] Although natural gas is not a resource of Puerto Rico, it is included as a major source of hydrocarbon raw materials for the various products to be discussed. The reason for its inclusion is that natural gas is available in other regions which are or may be competitive with Puerto Rico.

Consideration of the many possible combinations of processes and commodities, not all of which can be listed in Figure 1, raises the problem of identifying that complex which can achieve the

[2] For example, a glance at the top of Figure 1 indicates that methane can be converted into hydrogen, and this in successive stages into nitric acid, urea, and ultimately fertilizer. Methane also yields methanol which, via formaldehyde, yields plastics. Or methane can be converted into HCN and acetylene and ultimately into the synthetic fibers, Orlon, Dynel, and Acrilan. Or acrylonitrile can be produced from acetylene, which in turn may be produced from ethane, ethylene, propane, or propylene. Acetylene, however, may be used to produce acetic acid, which leads to rayon, or vinyl chloride, which can be processed into polyvinyl chloride and plastics. Or the ethylene can be converted into any number of products and ultimately into plastics, synthetic rubber, antiknock fluid, synthetic fibers, antifreeze, detergents, and explosives. Propane converted into propylene can also be used for the production of many of these end products or can enter into LPG, a high-grade liquid fuel. The naphthenes yield an interesting group of products of which the most familiar are nylon, synthetic rubber, plastics, paints, insecticides, and synthetic fibers.

most advantage in a Puerto Rico location. A first step might be an examination of the list of products at the extreme right of the figure. It seems logical that, of any group of products that utilize the same raw material or intermediate commodity, that one which also utilizes to the greatest degree the type of labor available in Puerto Rico is likely to show greatest advantage for a Puerto Rico location, *ceteris paribus*. Of the various products listed, synthetic fibers generally tend to utilize such labor to the greatest degree. Thus an industrial complex in Puerto Rico that included synthetic fiber production would take advantage of the first resource—an abundance of moderately skilled and cheap labor.

If to synthetic fiber production is linked oil refining, the resulting complex could gain from Puerto Rico's second resource—favorable location with respect to Venezuelan oil. (The raw materials for synthetic fibers can be obtained from petroleum refinery byproducts.) This combination results in additional gains on two scores: (1) Refinery by-products are upgraded, and (2) transport costs on chemical intermediates are eliminated. Lastly, both refinery products and synthetic fiber can be exported to the United States duty-free, thus taking advantage of Puerto Rico's third important resource—access to mainland markets.

The investigator must not be content, however, with a study of export potentials of a region in order to identify meaningful industrial opportunities. Important potentials can often be discovered by means of a survey of current as well as possible future imports of a region, although care must be taken to avoid the mercantilistic view that any reduction in imports is desirable for a region. A survey of the various imports into Puerto Rico indicates a very heavy inflow of fertilizer.[3] Since transport cost of fertilizer is a locationally important variable, and since nitrogen fertilizers can be based on refinery by-products, thus upgrading refinery by-products, it seems appropriate to include fertilizer production as a basic element in a complex to be investigated. This step is all the more reasonable in view of the similar production characteristics of fertilizer and fiber intermediates and in view of the large growth of fertilizer consumption that is to be expected in Puerto Rico over the years.

In sum, it can be said that, in considering the possibilities for

---

[3] See "United States Trade in Merchandise and Gold and Silver with United States Territories and Possessions," *Census Report FT800,* U. S. Department of Commerce, Bureau of the Census, Washington, 1953; and Perloff, *op. cit.,* pp. 135-144.

30 INDUSTRIAL COMPLEX ANALYSIS

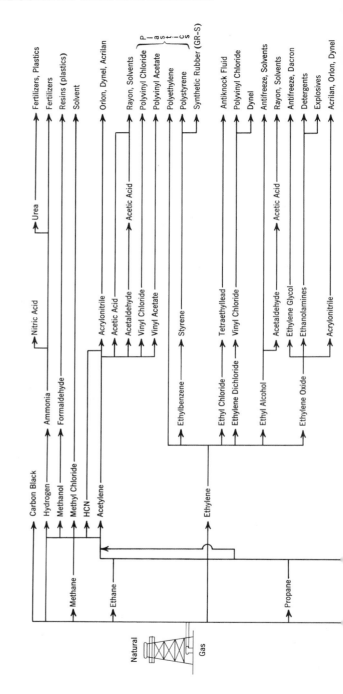

CHOICE OF RELEVANT INDUSTRIAL COMPLEXES 31

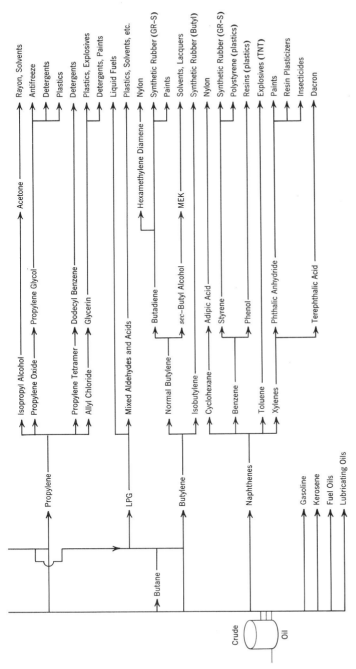

Figure 1. Flow sheet of principal petrochemical raw materials, intermediates, and end products.

32 INDUSTRIAL COMPLEX ANALYSIS

economic development in Puerto Rico, *industrial complexes* involving petroleum refinery, petrochemical, synthetic fiber, and fertilizer processes appear to be promising. Therefore the problem is reduced to the identification of one or several of these complexes which would be among the most advantageous ones for Puerto Rico relative to competing mainland locations.

ACTIVITIES EXCLUDED

At this point, however, a few remarks ought to be made regarding the exclusion from the complex of promising activities, such as electronics; textile spinning, weaving, and finishing; and chemurgic processes.

1. *Electronics* is typical of a class of activities that hold promise for Puerto Rico, but that have not been probed. These activities are for all practical purposes independent of the hydrocarbon-fiber complexes to be examined and can be analyzed separately. It is to be noted that all activities within these hydrocarbon-fiber complexes are connected by a chain of inputs and outputs. Electronics, on the other hand, has no significant interconnections of this kind with activities in these complexes.[4]

---

[4] Incidentally, there is little, if any, significant interdependence of electronics and such activities with respect to general supply–demand relationships. Unskilled and semiskilled labor is assumed in this study to be available at a fixed price; and the pool of such labor is assumed to be sufficient in size to meet the requirements of several basic industries, even when local multiplier effects are taken into account. In contrast, skilled labor is a *common* input in limited local supply; but in this study it is conservatively postulated that all skilled labor will be recruited from the mainland. Finally, both a hydrocarbon-fiber complex and an electronics industry would produce principally for export markets and would therefore be largely unaffected by local income changes induced by each.

The only exception to this last statement relates to the possibility of increased demand for fertilizer which would tend to result from significant increase in demand for locally produced agricultural products, which in turn might be associated with increase in local income due to electronics production. However, we shall retain the implicit assumption of the text, namely, that the demand for fertilizer is inelastic with income. Such an assumption is purely formal and applicable only to the present analysis. The demand for fertilizer can be posited to continue its sharp yearly rise observed in recent years. This rise is clearly not independent of income changes, but at the same time it depends strongly on other factors (social changes, educational progress) which may safely be judged to make it irreversible. Thus, it would seem that fertilizer demand can be better projected from a study of time series than from any conjecture about its income elasticity.

2. *Textile spinning, weaving, and finishing,* and the production of other labor-intensive end products based on chemical raw materials are a group of processes typical of a second class of activities omitted from consideration. These processes could obtain some of their various alternative raw materials from the petrochemical and the synthetic fiber operations, and could thus profit as activities within the complex. Such activities as these have a clear advantage in Puerto Rico. They could also upgrade the chemical and fiber products of the complex. The principal interest of the present study, however, lies in the evaluation of a set of strongly interrelated processes which at the same time are more basic, and for which the profitability for Puerto Rico is not clear at the start. The inclusion of clearly profitable activities would tend to conceal the advantages or disadvantages of the basic complex alone. It is evident that any complex that excludes, for example, textile processing activities and yet shows profitability for Puerto Rico would be even more favorable if these activities are included.[5]

TYPE OF COMPLEX TO BE EVALUATED

Apart from the problem of determining the types of activities to be included in a complex, there is a second methodological problem which relates to the choice of a group of individual complexes for comparison.

A complex is a set of activities at a specific location which are linked by certain technical and production interrelations. These interrelations may be such that a spatial juxtaposition of the activities in the complex leads to substantial economies. If so, a realistic locational or regional analysis must evaluate for each potential location the *entire* complex. Misleading or at best inconclusive results tend to emerge if each or any of the activities is analyzed independently of the others.

On the other hand, a set of activities may be technologically or otherwise interrelated and yet not be subject to the spatial inter-

---

[5] Some analysts, however, might include these activities in a complex because they judge that the resulting integration and other types of economies not explicitly considered in our analysis would be substantial. This would especially be so if negative conclusions regarding Puerto Rico were reached on the basis of the exclusion of these economies.

dependence just described. For example, a group of chemical activities might represent the various raw-material and intermediate production stages on the way to a final plastic product, and yet be subject to few if any economies stemming from spatial juxtaposition alone. It is true that considerations of regional differences in transport and other costs might lead to the establishment of a "complex" of some or all of such activities at a given location. It is also true that, starting from the basic raw material, each successive stage or activity would constitute at least part of the "market" for the immediately preceding stage, and hence the various activities to some extent would be locationally interdependent. However, the production interrelationships connecting such activities are not dependent upon the activities being spatially adjacent. Essentially the activities are spatially independent. For this reason, a separate locational analysis could be made of each activity. The picture of an over-all complex at each location could be built up in several steps. First, an analysis of the basic raw-material processing activity might be investigated; upon the results of this analysis, a study of the second-stage activity might be pursued; and so forth. Each such analysis would resemble an individual industry comparative-cost study; it would evaluate interregional differences in transport costs; in labor, power, and fuel costs; in economies of scale; etc. Clearly, different complexes of activities could develop. For a given market, all the activities might be included in one giant integrated complex concentrated at a single location; or these activities might be geographically split in any one of a number of ways, and give rise to several types of smaller complexes at several locations.

The locational analysis, of course, need not be built up by analyzing individual activities as just described. At the outset a locational comparison of the *entire* complex of activities in one region vis-a-vis other regions could be made, and this would be followed by further comparisons relating to the various possible split-location patterns. These analyses would consist of the same sort of interregional cost and revenue evaluation mentioned above.

It is unlikely that technically related activities would exhibit complete spatial independence. There are likely to be at least some complexes of two or more of such activities which among themselves exhibit strong interrelations leading to major economies of spatial integration. The existence of such complexes reduces the number of split-location possibilities to be analyzed, since the larger the number of activities in spatially interdependent complexes, the fewer the number of meaningful split-location patterns. At the extreme, spatial interdependence of all activities results in the type of complex already mentioned, for which all activities as a group must be evaluated for each possible location.

The economies that stem from geographic juxtaposition of activities are often difficult to quantify. This difficulty does not constitute a serious obstacle to analysis if it can be assumed with confidence that such economies are substantial enough to rule out any split-location pattern. In such a case it can usually be further postulated that such economies are equal in magnitude at all feasible locations. Thus it is sufficient for one to pursue a locational comparison of the full complex, using only data on regional differences in production and transport costs and in product revenues; such data are more easily obtainable.

A more difficult situation arises if economies of spatial integration exist but appear modest, and if locational analysis of regional cost and revenue differences indicates that a split-location pattern is most favorable. This situation requires direct comparison of the economies of spatial integration with the savings stemming from split location. If the former resist even rough and approximate quantification, the analysis can only identify the more favorable alternatives; it cannot choose among them.

As described above, the present study is concerned with the feasibility of an oil refinery–petrochemical–synthetic fiber complex in Puerto Rico. Many of the intermediate materials and products of such a complex can be easily transported by water and overland. Thus, there emerges the possibility of a split-location pattern involving complexes (or individual activities) both in Puerto Rico and on the mainland. However, judgment indicates and technical experience has demonstrated that the technically interrelated activities considered in this study tend to be subject to substantial, though largely nonquantifiable, economies of spatial integration. Among the many advantages, tangible and intangible, that a complex involving the full spatial integration of all activities enjoys compared to a split-location pattern involving two or more smaller complexes are (1) for underdeveloped areas the growth of pools of efficient labor and of management skills, (2) savings in indirect production costs, (3) finer articulation of production among the several stages, (4) possible scale advantages in nonbattery limits facilities, (5) more effective quality control. All these factors would be of relevance to a large integrated firm that is considering erecting new plants in Puerto Rico or on the mainland, or in both places. In addition, there are social welfare considerations which relate to the differential impacts of different size and type of complexes upon entrepreneurial attitudes, savings and investment habits, rate of industrialization, and basic development processes. These latter factors are significant when the problem is viewed by an economic development agency, or in general by a body politic; especially so in the case of Puerto Rico.

Given this background of considerations, the authors of the present study judged it most desirable to develop first and most comprehensively the locational comparison for *full* complexes. The analysis to be presented is limited for the most part to the relatively objective factors of locational importance; there is no evaluation of the intangible forces of spatial integration. However, in addition to the analysis of full programs, various possible split-location patterns are examined, partly in order to derive some conception of the minimum magnitude of economies of spatial integration that are necessary to overcome possible locational advantages of otherwise desirable split-location patterns.

BASIC POSTULATES. PLAN OF THE STUDY

An advantage for a Puerto Rico complex is defined to exist if its net revenue is greater than that of a similar, efficiently located and operated mainland complex. This net revenue could be greater if the Puerto Rico location could produce and deliver at a lower total cost a set of final outputs, or if for the same total cost it could obtain from sales to diverse markets a greater total revenue, or both. It is assumed that all inputs can be purchased and all outputs can be sold at fixed, determined prices, but that in Puerto Rico some of the potential activities are subject to rigid limits. It is further assumed that all locational decisions are made to maximize long-run profits; all short-run barriers and resistances are ignored.

Chapter 3 consists of the presentation and discussion of the various individual refinery, petrochemical, and synthetic fiber activities which are relevant for the general type of complex envisioned. Shown for each activity is an appropriate set of inputs and outputs, for which the sources of data and methods of estimation are discussed in Appendix A.

Chapter 4 sets up a number of specific programs (complexes) considered to be possible and meaningful for Puerto Rico. Each program is a different combination of individual refinery, petrochemical, and synthetic fiber activities. The number of possible

CHOICE OF RELEVANT INDUSTRIAL COMPLEXES 37

combinations is virtually limitless, but obviously only a finite number can be considered. The process of selection is discussed in detail.

Chapters 5 and 6 begin the evaluation in dollars and cents terms of the locational advantage of Puerto Rico for each selected complex. Chapter 5 contains a detailed derivation of individual commodity-price differences between Puerto Rico and the United States mainland. Chapter 6 develops different computational procedures and programming techniques whereby the commodity-price differences are employed to calculate the preliminary locational advantages or disadvantages of the selected programs. The virtues and limitations of the different techniques are discussed. Also in Chapter 6 several "short" programs are investigated, and their preliminary locational advantages or disadvantages in Puerto Rico are evaluated. These programs comprise fewer activities than the full-length programs.

In order to include an even wider range of results for comparison, preliminary locational advantages or disadvantages of all the full programs and short programs in Puerto Rico are calculated in accordance with a number of sets of price differences which vary from those derived in Chapter 5. The variations occur in the price differences postulated for crude oil, refinery products, chemical labor, and textile labor. The findings are presented in the relevant sections of Chapter 6.

The second major stage of the analysis concerns the adjustment of the preliminary locational-advantage magnitude for each selected complex in order to allow for differential profitabilities. The necessity for such adjustment stems from the fact that in actuality there are likely to be significant nonidentities between a Puerto Rico complex and the most efficient comparable mainland operation.[6] Chapter 7 explains and evaluates the major considerations and the methodology involved in the treatment of differential

---

[6] Actually, the distinction between differential profitabilities and ordinary price (revenue and cost) differences in an empirical problem is not quite as sharp as has just been implied. In a number of cases, it is possible to predict that a given activity, whenever it occurs in a program in Puerto Rico, will always substitute for another given activity on the mainland. For example, whenever ammonia is made from fuel oil in Puerto Rico, a comparable complex located at the Gulf Coast would always manufacture ammonia from fuel gas. In this instance, if the Btu's of fuel are assumed to be the basic inputs, the Puerto Rican and the mainland activities can be considered identical, since both processes use virtually the same amount of Btu's. Thus it is possible to associate with the fuel oil–ammonia activity a price difference that includes the differential profitability correction.

profitability corrections. Chapter 8 details the calculation of such corrections for the selected Puerto Rico complexes involving the production of nylon fiber. Chapter 9 presents in more summary fashion an account of the differential profitability corrections for the complexes involving Dacron, Orlon, and Dynel fiber production.

The initial sections of Chapter 10 contain summary comments on the differential profitability corrections applicable to the full programs. Then similar corrections are applied to the selected short programs. In the latter part of the chapter, differential profitability corrections are applied to the preliminary results obtained in Chapter 6 on the basis of different sets of commodity-price differences. After an evaluation of the probable effects of certain modifications and extensions of the differential profitability analysis (e.g., different assumptions regarding the nature and extent of demand for Puerto Rico petrochemical production), Chapter 10 closes with a final summary of findings and a statement of over-all conclusions.

In order to arrive at the "best" complex from the point of view of Puerto Rico, the general analysis could have been broadened to include the definition of a social goal and the provision for a variable tax-subsidy pattern. However, this extremely difficult extension was not attempted. For the most part the analysis pertains to a locational decision by an oligopolist. *If* intangibles can be ignored, the complex that has associated with it the maximum adjusted locational advantage may be accepted as designating the most desirable program. Where intangibles cannot be ignored the adjusted locational-advantage values associated with the different alternative complexes considered may be regarded as essential background information for decision making by a large integrated firm.[7] Also, these values can be considered essential materials for the appraisal of alternate development programs from the point of view of insular welfare. In such an appraisal, they will need to be contrasted with the subjective-type advantages of these several programs.

---

[7] For the firm, among the more important intangibles that are not explored in this study are the effect of distance and time on customer service and new sales, the effect of project timing on share of market, the community spirit and climate at a new location as evaluated by management, the potential labor difficulties as estimated by management, the accessibility of research and staff department facilities, the possibilities of further expansion in a region, and the possible effects of legislative changes.

Chapter 3

The Individual Production Activities: Inputs and Outputs

In the previous chapter were stated the various considerations governing the choice of general types of industrial complexes to be investigated. The objective of the study was more precisely defined, and the most promising lines of attack were identified. The point was reached from which the study could proceed with the analysis of specific industrial complexes, defined in terms of particular processes, product mixes, and levels of output.

The first step in such analysis centers around the collection and processing of a basic set of technical engineering (input-output) data. The problems associated with this step are discussed in this chapter.

THE BASIC ACTIVITY MATRIX

First it is desirable to set down the possible activities or processes that might characterize a refinery–petrochemical–synthetic fiber complex. A listing of such activities appears at the heads of the columns of Table 3. Excluded from the table are many activities which would labor under diseconomies of scale in Puerto Rico, or for which feasible and economic operation has not been as yet

TABLE 3—PART I
ANNUAL INPUTS AND OUTPUTS FOR SELECTED OIL REFINERY,

		Oil Refinery, Prototype #1	Oil Refinery, Prototype #2	Oil Refinery, Prototype #3	Oil Refinery, Prototype #4	Oil Refinery, Prototype #5
		1	2	3	4	5
1	Crude Oil MMbbl	− 9.428	− 9.428	− 9.428	− 9.428	− 9.428
2	Gasoline, straight-run MMbbl	+ 2.074	+ 2.074	+ 1.300	+ 1.300	+ 0.942
3	Gasoline, cracked MMbbl	+ 1.484	+ 2.226	+ 1.484	+ 2.226	+ 1.737
4	Gasoline, reformed MMbbl	....	....	+ 1.486	+ 1.486	+ 1.486
5	Gasoline, polymerized MMbbl	+ 0.219	+ 0.328	+ 0.306	+ 0.415	+ 0.636
6	Naphtha, MMbbl	+ 0.660	+ 0.660	....	....	....
7	Kerosene, MMbbl	+ 0.943	+ 0.707	+ 0.943	+ 0.707	+ 0.717
8	Diesel Oil MMbbl	+ 1.414	....	+ 1.414	+ 0.896	+ 1.075
9	Gas Oil MMbbl	....	....	+ 0.518	....	....
10	Cycle Oil MMbbl	+ 1.320	+ 1.980	+ 1.320	+ 1.980	+ 1.544
11	Heavy Residual MMbbl	+ 0.943	+ 0.943	....	....	....
12	Coke and Carbon 10XMM lbs	....	....	+ 4.033	+ 4.033	+ 26.810
13	L.P.G. 10XMMlb	+ 6.860	+ 10.280	+ 11.620	+ 15.050	+ 15.611
14	Hydrogen MMlb	+ 0.950	+ 1.430	+ 8.500	+ 8.900	+ 9.860
15	Methane MMlb	+ 12.780	+ 19.180	+ 28.460	+ 34.860	+ 70.810
16	Ethylene (mixed) MMlb	+ 6.510	+ 9.760	+ 14.150	+ 17.410	+ 41.000
17	Ethane (mixed) MMlb	+ 9.930	+ 14.900	+ 27.280	+ 32.250	+ 65.060
18	Propylene MMlb	+ 3.630	+ 5.430	+ 5.770	+ 7.580	+ 13.570
19	Propane MMlb	+ 2.150	+ 3.230	+ 4.010	+ 5.080	+ 5.270
22	Pure Ethylene MMlb	....	....	....	....	....
23	Pure Ethane MMlb	....	....	....	....	....
24	Steam MMMlb	− 0.801	− 1.024	− 1.179	− 1.402	− 1.805
25	Power MMKWH	− 2.511	− 3.411	− 3.168	− 3.999	− 5.220
26	Fuel 10XMMMBTU	−139.000	−157.000	−224.000	−242.000	−322.000
27	Salt MMlb	....	....	....	....	....
28	Caustic Soda MMlb	....	....	....	....	....
29	Chlorine MMlb	....	....	....	....	....
30	Hydrochloric Acid 100% MMlb	....	....	....	....	....
31	Limestone MMlb	....	....	....	....	....
32	Lime (hydrated) MMlb	....	....	....	....	....
33	Ethylene Oxide MMlb	....	....	....	....	....
34	Nitrogen MMlb	....	....	....	....	....
35	Ethylene Glycol MMlb	....	....	....	....	....
36	Ethylene Dichloride MMlb	....	....	....	....	....
37	Acetylene MMlb	....	....	....	....	....

Sources: See text

demonstrated in the United States.[1] For example, the production of cyclohexane from refinery gasoline streams, and the separation of paraxylene from mixed xylenes are not included. Such production requires very large quantities of raw-material feedstocks rel-

---

[1] It should be observed, however, that the exclusion of certain processes because they are currently not in use in the United States might bias results, since it is not at all to be expected that the optimal technology of a highly industrialized country would best fit underdeveloped areas.

INDIVIDUAL ACTIVITIES: INPUTS AND OUTPUTS 41

TABLE 3—PART I
PETROCHEMICAL, AND SYNTHETIC FIBER ACTIVITIES

Oil Refinery, Prototype #6	Ethylene separation (Prototype #1)	Ethylene separation (Prototype #2)	Ethylene separation (Prototype #3)	Ethylene separation (Prototype #4)	Ethylene separation (Prototype #5)	Ethylene separation (Prototype #6)	Propane from LPG (dummy)	Ethylene from Ethane
6	7	8	9	10	11	12	13	14
− 9.428	....	....	....	....	....	....	....	....
+ 1.084	....	....	....	....	....	....	....	....
+ 2.226	....	....	....	....	....	....	....	....
+ 1.486	....	....	....	....	....	....	....	....
+ 0.328	+ 0.014	+ 0.021	+ 0.022	+ 0.029	+ 0.052	+ 0.021	....	....
....	....	....	....	....	....	....	....	....
+ 0.707	....	....	....	....	....	....	....	....
....	....	....	....	....	....	....	....	....
+ 1.980	....	....	....	....	....	....	....	....
+ 0.947	....	....	....	....	....	....	....	....
+ 14.110	+ 0.215	+ 0.323	+ 0.401	+ 0.508	+ 0.527	+ 0.469	− 1.000	+ 0.136
+ 8.480	....	....	....	....	....	....	....	+ 0.800
+ 22.630	....	....	....	....	....	....	....	+ 0.880
+ 9.090	− 6.020	− 9.030	−13.090	−16.100	−37.930	− 8.410	....	+ 0.810
+ 20.430	− 9.390	−14.080	−25.260	−30.190	−61.320	−19.140	....	+ 0.700
+ 5.450	− 3.630	− 5.430	− 5.770	− 7.580	−13.570	− 5.450	....	....
+ 4.690	− 2.150	− 3.230	− 4.010	− 5.080	− 5.270	− 4.690	+10.000	....
....	+ 6.020	+ 9.030	+13.090	+16.100	+37.930	+ 8.410	....	+10.000
....	+ 9.390	+14.080	+25.260	+30.190	+61.320	+19.140	....	−12.900
− 1.286	− 0.056	− 0.084	− 0.119	− 0.148	− 0.343	− 0.079	....	− 0.089
− 3.490	− 0.074	− 0.074	− 0.111	− 0.157	− 0.194	− 0.448	− 0.103	− 0.110
−216.000	....	....	....	....	....	....	....	− 7.680
....	....	....	....	....	....	....	....	....
....	....	....	....	....	....	....	....	....
....	....	....	....	....	....	....	....	....
....	....	....	....	....	....	....	....	....
....	....	....	....	....	....	....	....	....
....	....	....	....	....	....	....	....	....
....	....	....	....	....	....	....	....	....
....	....	....	....	....	....	....	....	....
....	....	....	....	....	....	....	....	....

ative to output quantity. At least several oil refinery operations (or a huge single operation) are required to furnish the necessary feedstock for an economic-size cyclohexane or paraxylene plant. Such a magnitude of refinery operation does not appear feasible for Puerto Rico. As another example, the production of ethylene from refinery gas oil fractions is excluded from the activities of Table 3. Under present market conditions, gas oil is much more profitably used to produce motor fuel.

At the extreme left of Table 3 are listed various commodities

INDUSTRIAL COMPLEX ANALYSIS

TABLE 3—Part I
ANNUAL INPUTS AND OUTPUTS FOR SELECTED OIL REFINERY,

	Ethylene from Propane	Chlorine by Electrolysis	Hydrochloric Acid from Chlorine	Hydrated Lime	Ethylene Oxide (Chlorhydrin)	Ethylene Oxide (Oxidation)
	15	16	17	18	19	20
1 Crude Oil MMbbl	....	....	....	....	....	....
2 Gasoline, straight-run MMbbl	....	....	....	....	....	....
3 Gasoline, cracked MMbbl	....	....	....	....	....	....
4 Gasoline, reformed MMbbl	....	....	....	....	....	....
5 Gasoline, polymerized MMbbl	....	....	....	....	....	....
6 Naphtha, MMbbl	....	....	....	....	....	....
7 Kerosene, MMbbl	....	....	....	....	....	....
8 Diesel Oil MMbbl	....	....	....	....	....	....
9 Gas Oil MMbbl	....	....	....	....	....	....
10 Cycle Oil MMbbl	....	....	....	....	....	....
11 Heavy Residual MMbbl	....	....	....	....	....	....
12 Coke and Carbon 10XMM lbs	....	....	....	....	....	....
13 L.P.G. 10XMMlb	+ 0.129	....	....	....	....	....
14 Hydrogen MMlb	+ 0.660	+ 0.290	- 0.350	....	....	....
15 Methane MMlb	+11.560	....	....	....	....	....
16 Ethylene (mixed) MMlb	+ 0.790	....	....	....	....	....
17 Ethane (mixed) MMlb	- 0.460	....	....	....	....	....
18 Propylene MMlb	+ 5.400	....	....	....	....	....
19 Propane MMlb	-29.800	....	....	....	....	....
22 Pure Ethylene MMlb	+10.000	....	....	....	- 7.520	-11.780
23 Pure Ethane MMlb	....	....	....	....	....	....
24 Steam MMMlb	- 0.119	- 0.005	....	....	- 0.064	- 0.079
25 Power MMKWH	- 0.560	-17.100	- 0.030	- 3.500	- 0.601	- 0.612
26 Fuel 10XMMMBTU	-20.660	....	....	- 3.500	....	- 2.851
27 Salt MMlb	....	-17.000	....	....	....	....
28 Caustic Soda MMlb	....	+11.250	....	....	....	....
29 Chlorine MMlb	....	+10.000	- 9.810	....	-16.760	....
30 Hydrochloric Acid 100% MMlb	....	+ 0.200	+10.000	....	....	....
31 Limestone MMlb	....	....	....	-11.430	....	....
32 Lime (hydrated) MMlb	....	....	....	+10.000	-13.380	....
33 Ethylene Oxide MMlb	....	....	....	....	+10.000	+10.000
34 Nitrogen MMlb	....	....	....	....	....	+68.000
35 Ethylene Glycol MMlb	....	....	....	....	....	....
36 Ethylene Dichloride MMlb	....	....	....	....	....	....
37 Acetylene MMlb	....	....	....	....	....	....

Sources: See text

which are either inputs or outputs, or both, of the diverse activities considered feasible.[2] In each column are recorded the amount of each commodity used as an input or output of the activity listed at the top, when that activity is operated at *unit level*. A unit level of operation may be defined for any activity as that level which

---

[2] Sources of the material are the following: activities 1-13, 57, and 66, see Appendix A; activities 14-17, 19-41, 43, 55, 56, see Walter Isard and Eugene W. Schooler, *Location Factors in the Petrochemical Industry*, U. S. Department of Commerce, Office of Technical Services, 1955, and unpublished materials; activities 44-54, 58-64, 67-73, Joseph Airov, *The Location of the Synthetic Fiber Industry*, John Wiley & Sons, 1959;

INDIVIDUAL ACTIVITIES: INPUTS AND OUTPUTS

TABLE 3—PART I
PETROCHEMICAL, AND SYNTHETIC FIBER ACTIVITIES

Ethyl-ene Glycol (Chlor-hydrin)	Ethyl-ene Glycol (Oxid-ation)	Ethyl-ene Di-(Chlor-ide)	Acetyl-ene from Methane (Wulff)	Acetyl-ene from Ethyl-ene (Wulff)	Acetyl-ene from Ethane (Wulff)	Acetyl-ene from Propyl-ene	Acetyl-ene from Propane (Wulff)
21	22	23	24	25	26	27	28
....	....	....	....	....	....	....	....
....	....	....	....	....	....	....	....
....	....	....	....	....	....	....	....
....	....	....	....	....	....	....	....
....	....	....	....	....	....	....	....
....	....	....	....	....	....	....	....
....	....	....	....	....	....	....	....
....	....	....	....	....	....	....	....
....	....	....	....	....	....	....	....
....	....	....	....	....	....	....	....
....	....	....	....	....	....	....	....
....	....	....	....	....	....	....	....
....	....	....	-34.200	....	....	....	....
....	....	....	....	-18.200	....	....	....
....	....	....	....	....	-19.500	....	....
....	....	....	....	....	....	-23.600	....
....	....	....	....	....	....	....	-24.700
- 5.920	- 8.300	- 2.950	....	....	....	....	....
- 0.088	- 0.103	- 0.001	- 0.245	- 0.130	- 0.140	- 0.170	- 0.179
- 0.750	- 0.800	- 0.020	- 2.060	- 1.270	- 1.270	- 0.750	- 0.750
....	- 2.010	....	+35.220	+ 7.940	+ 9.510	+12.140	+19.000
....	....	....	....	....	....	....	....
-13.190	....	- 7.210	....	....	....	....	....
....	....	....	....	....	....	....	....
-10.530	....	....	....	....	....	....	....
....	+68.000	....	....	....	....	....	....
+10.000	+10.000	....	....	....	....	....	....
....	....	+10.000	....	....	....	....	....
....	....	....	+10.000	+10.000	+10.000	+10.000	+10.000

activities 18, 42, R. N. Shreve, *The Chemical Process Industries,* McGraw-Hill Publishing Co., 1945, pp. 202 and 376; W. L. Faith, D. B. Keyes, and R. L. Clark, *Industrial Chemicals,* 2nd ed., John Wiley & Sons, 1957, pp. 484 and 743. (For sulfuric acid manufacture, only the contact process was considered, since this yields 100% acid, as required, while the chamber process yields dilute acid); activity 65, "How to Make Butadiene," *Chemical Engineering,* **61,** 306–309, (Sept. 1954); Rubber Producing Facilities Disposal Commission, *Government-Owned Synthetic Rubber Facility, PLANCOR 706, Lake Charles, Louisiana,* PBD-1, Washington, 1953; and private communications from several petroleum companies.

It is to be noted that for activities 58, 67, and 70, chemically pure hydrogen gas is required in the hydrogenation steps. In the present study, no separate activity for the production of chemically pure hydrogen is defined. It is assumed that the am-

INDUSTRIAL COMPLEX ANALYSIS

TABLE 3—PART II
ANNUAL INPUTS AND OUTPUTS FOR SELECTED OIL REFINERY

	Vinyl Chloride (Ethylene Dichloride)	Vinyl Chloride (Acetylene)	Ammonia from Hydrogen	Ammonia from Methane	Ammonia from Ethylene
	29	30	31	32	33
10 Cycle Oil MMbbl	....	....	....	....	....
11 Heavy Residual MMbbl	....	....	....	....	....
14 Hydrogen MMlb	....	....	- 2.000	....	....
15 Methane MMlb	....	....	....	- 5.500	....
16 Ethylene (mixed) MMlb	....	....	....	....	- 6.290
17 Ethane (mixed) MMlb	....	....	....	....	....
18 Propylene MMlb	....	....	....	....	....
19 Propane	....	....	....	....	....
24 Steam MMMlb	- 0.020	- 0.030	....	- 0.023	- 0.023
25 Power MMKWH	- 0.150	- 1.000	- 4.640	- 5.600	- 5.600
26 Fuel 10XMMMBTU	- 2.370	....	....	- 0.450	- 0.450
30 Hydrochloric Acid 100% MMlb	+ 6.230	- 6.900	....	....	....
33 Ethylene Oxide MMlb	....	....	....	....	....
35 Ethylene Glycol MMlb	....	....	....	....	....
36 Ethylene Dichloride MMlb	-17.210	....	....	....	....
37 Acetylene MMlb	....	- 4.300	....	....	....
38 Vinyl Chloride MMlb	+10.000	+10.000	....	....	....
39 Ammonia MMlb	....	....	+10.000	+10.000	+10.000
40 HCN MMlb	....	....	....	....	....
41 Acrylonitrile MMlb	....	....	....	....	....
42 Methanol MMlb	....	....	....	....	....
43 Sulphur MMlb	....	....	....	....	....
44 Sulphuric Acid MMlb	....	....	....	....	....
45 Nitric Acid MMlb	....	....	....	....	....
46 Paraxylene MMlb	....	....	....	....	....
47 Dimethyl Terephthalate MMlb	....	....	....	....	....
48 Dacron Polymer MMlb	....	....	....	....	....
49 Dacron Staple MMlb	....	....	....	....	....
50 Dacron Filament MMlb	....	....	....	....	....
51 Dynel Polymer MMlb	....	....	....	....	....
52 Acetone MMlb	....	....	....	....	....
53 Dynel Staple MMlb	....	....	....	....	....
54 Dynel Filament	....	....	....	....	....
55 Orlon Polymer MMlb	....	....	....	....	....
56 Dimethyl Formamide MMlb	....	....	....	....	....
57 Orlon Staple MMlb	....	....	....	....	....
58 Orlon Filament	....	....	....	....	....
59 Ammonium Nitrate MMlb	....	....	....	....	....
60 Urea MMlb	....	....	....	....	....
61 Carbon Dioxide MMlb	....	....	....	....	....

Sources: See text

yields on an annual basis the amount of primary product listed in the corresponding column. For most petrochemical and fiber activities this amount is 10 MM lb.

monia production process which produces chemically pure hydrogen as an intermediate product would simply be robbed of the required small amount of hydrogen. In Table 3, the hydrogen requirement has been listed in the row on mixed hydrogen; in later programming, however, an amount of ammonia chemically equivalent to the required pure hydrogen is scheduled as an input.

INDIVIDUAL ACTIVITIES: INPUTS AND OUTPUTS

TABLE 3—PART II
PETROCHEMICAL, AND SYNTHETIC FIBER ACTIVITIES

Ammonia from Ethane	Ammonia from Propylene	Ammonia from Propane	Ammonia from Heavy Residual	Ammonia from Cycle Oil	HCN from Methane and Ammonia	Acrylo-nitrile from Acetylene and HCN	Acrylo-nitrile from Ethylene Oxide	Sulphuric Acid from Sulphur
34	35	36	37	38	39	40	41	42
....	....	....	....	- 0.023	....	....	....	....
....	....	....	- 0.023	....	....	....	....	....
....	....	....	....	....	....	....	....	....
....	....	....	....	....	-11.800	....	....	....
....	....	....	....	....	....	....	....	....
- 5.780	....	....	....	....	....	....	....	....
....	- 6.420	....	....	....	....	....	....	....
....	....	- 6.050	....	....	....	....	....	....
- 0.023	- 0.023	- 0.023	- 0.050	- 0.050	- 0.019	- 0.208	- 0.084	+ 0.009
- 5.600	- 5.600	- 5.600	- 6.000	- 6.000	- 1.400	- 3.800	- 4.700	- 0.030
- 0.450	- 0.450	- 0.450	- 0.625	- 0.625	....	....	- 2.500	....
....	....	....	....	....	....	....	-10.200	....
....	....	....	....	....	....	....	....	....
....	....	....	....	....	....	....	....	....
....	....	....	....	....	....	- 6.600	....	....
+10.000	+10.000	+10.000	+10.000	+10.000	- 8.500	....	....	....
....	....	....	....	....	+10.000	- 6.600	- 5.100	....
....	....	....	....	....	....	+10.000	+10.000	....
....	....	....	....	....	....	....	....	- 3.440
....	....	....	....	....	....	....	....	+10.000
....	....	....	....	....	....	....	....	....
....	....	....	....	....	....	....	....	....
....	....	....	....	....	....	....	....	....
....	....	....	....	....	....	....	....	....
....	....	....	....	....	....	....	....	....
....	....	....	....	....	....	....	....	....
....	....	....	....	....	....	....	....	....
....	....	....	....	....	....	....	....	....
....	....	....	....	....	....	....	....	....
....	....	....	....	....	....	....	....	....
....	....	....	....	....	....	....	....	....

The first column of Table 3 describes the annual set of inputs and outputs of a refinery operation in Puerto Rico which includes a topping and vacuum flash unit, a fluid catalytic-cracking unit, a catalytic-polymerization unit, and a simple gas-separation plant. This combination is labeled oil refinery prototype 1. The first figure in column 1 is -9.428. This corresponds to the first row, which is labeled crude oil, million barrels. The negative sign before the number indicates that crude oil is an input. For a unit level of activity 1, thus 9.428 MM bbl of crude is required annually.

46 INDUSTRIAL COMPLEX ANALYSIS

TABLE 3—PART II
ANNUAL INPUTS AND OUTPUTS FOR SELECTED OIL REFINER*

		Nitric Acid from Ammonia	Di-methyl Terephthalate (air ox.)	Di-methyl Terehpthalate (nitric acid oxid.)	Dacron Polymer	Dacron Staple
		43	44	45	46	47
10	Cycle Oil MMbbl	....	....	....	....	....
11	Heavy Residual MMbbl	....	....	....	....	....
14	Hydrogen MMlb	....	....	....	....	....
15	Methane MMlb	....	....	....	....	....
16	Ethylene (mixed) MMlb	....	....	....	....	....
17	Ethane (mixed) MMlb	....	....	....	....	....
18	Propylene MMlb	....	....	....	....	....
19	Propane	....	- 0.030	- 0.095	- 0.060	- 0.500
24	Steam MMMlb	....	- 5.200	- 2.600	- 2.500	-12.000
25	Power MMKWH	- 1.200				
26	Fuel 10XMMMBTU	....	- 2.800	- 1.400	- 1.000	....
30	Hydrochloric Acid 100% MMlb	....	....	....	....	....
33	Ethylene Oxide MMlb	....	....	....	....	....
35	Ethylene Glycol MMlb	....	....	....	- 3.230	....
36	Ethylene Dichloride MMlb	....	....	....	....	....
37	Acetylene MMlb	....	....	....	....	....
38	Vinyl Chloride MMlb	....	....	....	....	....
39	Ammonia MMlb	- 2.860	....	....	....	....
40	HCN MMlb	....	....	....	....	....
41	Acrylonitrile MMlb	....	....	....	....	....
42	Methanol MMlb	....	- 4.000	- 4.000	+ 3.350	....
43	Sulphur MMlb	....	....	....	....	....
44	Sulphuric Acid MMlb	....	....	- 1.400	....	....
45	Nitric Acid MMlb	+10.000	....	-11.800	....	....
46	Paraxylene MMlb	....	- 6.800	- 5.900	....	....
47	Dimethyl Terephthalate MMlb	....	+10.000	+10.000	-10.000	....
48	Dacron Polymer MMlb	....	....	....	+10.000	-10.000
49	Dacron Staple MMlb	....	....	....	....	+10.000
50	Dacron Filament MMlb	....	....	....	....	....
51	Dynel Polymer MMlb	....	....	....	....	....
52	Acetone MMlb	....	....	....	....	....
53	Dynel Staple MMlb	....	....	....	....	....
54	Dynel Filament	....	....	....	....	....
55	Orlon Polymer MMlb	....	....	....	....	....
56	Dimethyl Formamide MMlb	....	....	....	....	....
57	Orlon Staple MMlb	....	....	....	....	....
58	Orlon Filament	....	....	....	....	....
59	Ammonium Nitrate MMlb	....	....	....	....	....
60	Urea MMlb	....	....	....	....	....
61	Carbon Dioxide MMlb	....	....	....	....	....

Sources: See text

The next figure in column 1 is +2.074. This signifies that associated with a unit level of activity 1 is an annual output of 2.074 MM bbl of straight run and coked gasoline. The third figure of column 1 indicates that 1.484 MM bbl of cracked gasoline is produced annually by activity 1 when it is operating at unit level. There is no figure in row 4 of column 1; no reformed gasoline is produced. And so forth. The plus sign always indicates an output; the neg-

INDIVIDUAL ACTIVITIES: INPUTS AND OUTPUTS 47

TABLE 3—Part II
PETROCHEMICAL, AND SYNTHETIC FIBER ACTIVITIES

Dacron Filament	Dynel Polymer	Dynel Staple	Dynel Filament	Orlon Polymer	Orlon Staple	Orlon Filament	Ammonium Nitrate from Ammonia	Urea from Ammonia
48	49	50	51	52	53	54	55	56
....	....	....	....	....	....	....	....	....
....	....	....	....	....	....	....	....	....
....	....	....	....	....	....	....	....	....
....	....	....	....	....	....	....	....	....
....	....	....	....	....	....	....	....	....
....	....	....	....	....	....	....	....	....
- 0.550	- 0.030	- 0.500	- 0.550	- 0.020	- 0.500	- 0.550	- 0.007	- 0.028
-15.000	- 1.200	-12.000	-15.000	- 1.500	-12.000	-15.000	- 0.170	- 0.340
....	....	....	....	....	....	....	....	....
....	....	....	....	....	....	....	....	....
....	....	....	....	....	....	....	....	....
....	....	....	....	....	....	....	....	....
....	- 6.300	....	....	....	....	....	....	....
....	....	....	....	....	....	....	- 2.380	- 5.800
....	....	....	....	....	....	....	....	....
....	- 4.200	....	....	-10.500	....	....	....	....
....	....	....	....	....	....	....	....	....
....	....	....	....	....	....	....	....	....
....	....	....	....	....	....	....	- 7.630	....
....	....	....	....	....	....	....	....	....
-10.000	....	....	....	....	....	....	....	....
....	....	....	....	....	....	....	....	....
+10.000	....	....	....	....	....	....	....	....
....	+10.000	-10.000	-10.000	....	....	....	....	....
....	....	- 3.000	- 3.000	....	....	....	....	....
....	....	+10.000	....	....	....	....	....	....
....	....	....	+10.000	....	....	....	....	....
....	....	....	....	+10.000	-10.000	-10.000	....	....
....	....	....	....	....	- 2.300	- 2.300	....	....
....	....	....	....	....	+10.000	....	....	....
....	....	....	....	....	....	+10.000	....	....
....	....	....	....	....	....	....	+10.000	....
....	....	....	....	....	....	....	....	+10.000
....	....	....	....	....	....	....	....	- 7.500

ative sign, an input. Further down column 1 appear the usual utilities: steam, power, and fuel.

Column 2 refers to the inputs and outputs associated with a unit level of operation of oil refinery prototype 2. This activity differs from the first activity in that it embodies more extensive cracking operations. Preliminary considerations indicate that the two degrees of cracking associated with activities 1 and 2 could be

48 INDUSTRIAL COMPLEX ANALYSIS

TABLE 3—PART III
ANNUAL INPUTS AND OUTPUTS FOR SELECTED OIL REFINERY

			Benzene from Refinery	Cyclohexane from Benzene	Adipic acid from Cyclohexane (Nit.)	Adipic acid from Cyclohexane (Air)	Furfural from Bagasse	Tetrahydrofuran from Furfural
			57	58	59	60	61	62
2	Gasoline, STR Run,	MMbbl	- 1.084					
4	Gasoline, Reformed,	MMbbl	+ 0.987					
5	Gasoline, Polymerized,	MMbbl						
14	Hydrogen,	MMlb		- 0.611				
21	Butane,	MMlb						
24	Steam,	MMlb	- 0.380	- 0.105	- 0.129	- 0.143	- 0.400	- 0.13
25	Power,	MMKwh	- 0.122	- 1.130	- 3.400	- 2.300	- 0.250	- 1.25
26	Fuel,	10MMBtu			- 0.894	- 3.062		- 1.4
28	Caustic Soda,	MMlb						
29	Chlorine,	MMlb						
30	Hydrochl. Acid, 100%	MMlb						
31	Limestone,	MMlb						
34	Nitrogen,	MMlb						
39	Ammonia,	MMlb						
40	HCN,	MMlb						
44	Sulfuric Acid,	MMlb					- 2.250	
45	Nitric Acid,	MMlb			-12.370			
62	Benzene,	MMlb	+24.360	- 7.970				
63	Cyclohexane (85%),	MMlb		+10.000	-12.550	-16.940		
64	Cyclohexanone,	MMlb						
65	Adipic Acid,	MMlb			+10.000	+10.000		
66	Bagasse,	MMlb					-220.000	
67	Furfural,	MMlb					+10.000	-16.9
68	Tetrahydrofuran,	MMlb						+10.0
69	Sodium Cyanide,	MMlb						
70	Adiponitrile,	MMlb						
71	Butadiene,	MMlb						
72	Ammonium Sulfate,	MMlb						
73	Hexamethylene Diamine,	MMlb						
74	Nylon Salt,	MMlb						
75	Nylon Staple,	MMlb						
76	Nylon Filament,	MMlb						

Sources: See text

relevant for Puerto Rico. Column 3 represents inputs and outputs for the operation at unit level of oil refinery prototype 3. In addition to the units contained in prototype 1, prototype 3 includes a coking unit and a re-forming unit. Column 4 differs from column 3 only in that cracking is carried on to a greater extent.

There are any number of refinery prototypes that could be considered. However, Table 3 presents input and output data for only

INDIVIDUAL ACTIVITIES: INPUTS AND OUTPUTS 49

TABLE 3—PART III
PETROCHEMICAL, AND SYNTHETIC FIBER ACTIVITIES

Sodium Cyanide from Hydr. Cyanide and Caustic	Adiponitrile from Tetrahydrofuran	Butadiene from Butane	Butadiene Refinery	Adiponitrile from Butadiene	Adiponitrile from Adipic acid	Adipic acid from Adiponitrile	Hexamethylene Diamine from Adiponitrile	Nylon Salt	Nylon Staple	Nylon Filament
63	64	65	66	67	68	69	70	71	72	73
....	....	....	....	....	....	....	....	....	....	....
....	....	....	....	....	....	....	....	....	....	....
....	....	....	- 0.054	....	....	....	....	....	....	....
....	....	....	....	- 0.195	....	....	- 0.731	....	....	....
....	....	-18.200	....	....	....	....	....	....	....	....
- 0.015	- 0.037	- 0.460	- 0.188	- 0.146	- 0.025	- 0.027	- 0.105	- 0.090	- 0.505	- 0.555
- 0.300	- 0.460	- 1.620	- 2.060	- 1.500	- 1.500	- 0.300	- 1.130	- 1.100	-13.000	-16.000
....	-2.232	-33.050	-19.159	- 1.230	- 2.090	....	....	- 2.000	....	....
-8.200	....	....	....	....	....	....	....	....	....	....
....	....	....	....	- 9.100	....	....	....	....	....	....
....	- 8.890	....	....	+ 1.000	....	....	....	....	....	....
....	....	....	....	-10.100	....	....	....	....	....	....
....	....	....	....	....	....	....	....	....	- 2.200	- 2.200
....	....	....	....	- 3.400	....	....	....	....	....	....
- 5.500	....	....	....	- 5.500	....	....	....	....	....	....
....	....	....	....	....	- 7.450	....	....	....	....	....
....	....	....	....	....	....	....	....	....	....	....
....	....	....	....	....	-18.210	+10.000	....	- 6.460	....	....
....	- 8.770	....	....	....	....	....	....	....	....	....
+10.000	- 9.640	....	....	....	....	....	....	....	....	....
....	+10.000	....	....	+10.000	+10.000	- 8.210	-10.230	....	....	....
....	....	+10.000	+10.000	- 7.400	....	+10.000	....	....	....	....
....	....	....	....	....	....	+10.000	- 5.130	....	....	....
....	....	....	....	....	....	....	+10.000	-10.000	-10.000
....	....	....	....	....	....	....	....	+10.000	....
....	....	....	....	....	....	....	....	....	+10.000

the six that were considered most relevant. This limitation helps to restrict Table 3 to a reasonable size.[3]

The next six columns relate to the inputs and outputs for gas

---

[3] In fact, because of the overwhelming number of industrial complexes that are possible, it was found desirable to consider only one type of oil refinery operation for Puerto Rico, namely that one denoted as prototype 4. From preliminary considera-

separation processes. These processes are necessary when it is expressly desired to separate ethylene and ethane from the mixed refinery gases for later use in various petrochemical activities. The next columns refer in succession to the cracking of ethane to obtain ethylene, the cracking of propane to obtain ethylene, the electrolysis of salt to yield chlorine, the production of hydrochloric acid from chlorine, the production of hydrated lime, the production of ethylene oxide from ethylene via the chlorhydrin process, etc.

In short, each column refers to a meaningful process or activity. So far as possible these activities are listed in the order in which they occur, beginning with the refining of crude oil. The sets of outputs and inputs associated with these activities (as recorded in Table 3) comprise the basic structural matrix for the ensuing analysis.

The inputs and outputs of Table 3 are assumed to be invariant with respect to location. It is of course recognized that actual productive activities may differ somewhat between the continental United States and Puerto Rico. For example, differences in climate or in the availability of ground water could cause differences in process design. However, variations such as these, as well as others, between locations may safely be considered minor and disregarded.

COMMODITIES OMITTED FROM THE BASIC MATRIX

Certain types of commodities are omitted from Table 3. One type consists of commodities such as oxygen and cooling water for which the price or cost difference between a Puerto Rican and a mainland location is negligible, if not zero. Such commodities, whether inputs or outputs, do not give rise to locational advantage or disadvantage so long as identical amounts are consumed or produced by the mainland and the Puerto Rico complexes being compared. This situation of course obtains in the initial stage of the analysis, in which identical complexes, except for location, are con-

---

tion, this prototype was judged to be the most favorable for Puerto Rico. After a significant amount of analysis had been conducted, there was no indication that the consideration of other prototypes would have improved the "profitability" position of Puerto Rico, or would have uncovered economies for mainland operation that are not already embodied in the study. For further discussion see pp. 147–148.

sidered. The omission of these commodities from Table 3 means that their amounts, as inputs or outputs of any activity, need not be determined. The result is a significant saving of computation time. It is true, however, that at a later stage of analysis, when nonidentical mainland and Puerto Rico complexes are compared, many of the omitted commodities must be considered and their amounts calculated. Such is the case, for example, if the amount of a given commodity used as an input in a mainland complex differs from the amount used in a comparable Puerto Rico complex.

Also omitted from Table 3 are those commodities whose physical weights (either as inputs or outputs) are minor when compared to the weights of the principal inputs and outputs of an activity. Such items as platinum catalysts fall into this category. Though their use may contribute substantially to the cost of an activity, the difference in cost between Puerto Rico and the mainland attributable to the use of such commodities is necessarily insignificant, since such differences cannot exceed transport costs between the two locations.[4]

For the productive activities of Table 3, there is considerable empirical evidence to indicate that the recorded inputs and outputs, such as feedstocks, products, and utilities, as well as other inputs such as catalysts and auxiliary chemicals, tend to vary in an essentially linear fashion with the scales of the processes.[5] For example, if the level of activity of oil refinery prototype 1 is doubled, all the figures of column 1 (which relate to a unit level of activity) can be doubled to obtain the approximate set of annual inputs and outputs appropriate to the new activity level. Or, if ethylene is to be produced from ethane at a level of activity of 3, the inputs and outputs recorded in column 14 can be tripled to obtain reasonably accurate information on total inputs and outputs.

However, certain inputs do not vary with scale in this simple linear fashion. Such inputs constitute another major type of commodity omitted from Table 3. For the purposes of the present study the only significant inputs of this type are capital and direct labor services, plus certain indirect cost elements based on these

---

[4] In the present study, it is assumed that transport rates applicable to such commodities are not exorbitantly high.

[5] See Isard and Schooler, *op. cit.,* pp. 21–23 and references there cited; also John Robert Lindsay, *The Location of Oil Refining in the United States,* unpublished Ph.D dissertation, Harvard University, 1954, p. 281 and references there cited.

inputs.[6] Each of these types of inputs will now be discussed in more detail.

CAPITAL INPUTS

It has been established from operating experience that capital inputs (i.e., investment costs of plant and equipment) in the types of activities represented in Table 3 do not vary linearly with activity scale. For the most part this situation reflects a number of physical or technical relationships. For example, doubling the capacity of a reaction vessel does not entail doubling the quantity (and hence the cost) of metal required in its construction. There is considerable empirical evidence that, over significant capacity ranges, capital (investment) costs of complete plants, as well as of individual equipment items, tend to vary in accordance with a fractional power law of the type

$$I = I_k \left(\frac{O}{O_k}\right)^\alpha \qquad (0 < \alpha \leq 1)$$

where I, which is to be estimated, denotes investment in building, equipment, and other capital items, for a plant under consideration with a projected capacity of O, and where I_k represents known investment for a plant of capacity O_k producing the same commodity.[7] The exponent α is usually designated the "plant factor." It varies

---

[6] Capital charges are omitted from Table 3 for an additional reason. It is assumed provisionally that the analysis pertains to a large integrated firm which might operate individual activities either on the mainland or in Puerto Rico. Such a firm would almost certainly draw investment funds from the capital market of the United States. Accordingly, it is assumed in this study that the cost of capital funds will be invariant with respect to the firm's locational decisions. Admittedly, the credit sources of the firm might judge the risk of a Puerto Rico location somewhat higher than that of a mainland location, and might offer less favorable credit terms for a Puerto Rico investment project. Likewise, the required return on equity capital might be set somewhat higher by the firm, in order to balance an assumed addition to risk. As an offset, however, the Puerto Rico government might make available funds on a more favorable basis than would be attainable in the mainland capital markets. Despite these considerations, the assumption of identical capital costs on the mainland and in Puerto Rico is deemed justifiable for this study.

[7] Cost versus capacity correlations are available for a wide variety of process equipment and machinery, ranging from pipes to heat exchangers and from pumps to vibrating screens. Such information appears regularly in petroleum refining and chemical engineering trade periodicals, e.g., *Oil and Gas Journal* and *Chemical Engineering*. The validity of fractional power functions with respect to complete plant costs is developed in Cecil H. Chilton, "Six-Tenths Factor Applies to Complete Plant Costs," *Chemical Engineering*, **57**, 112–114 (Apr. 1950).

INDIVIDUAL ACTIVITIES: INPUTS AND OUTPUTS 53

from activity to activity, and applies when for any given activity both O and O_k are within the meaningful capacity range discussed below.

Neither the size of an individual process unit nor the capacity of a whole plant can be increased indefinitely. Practical engineering, transportation, and marketing considerations, as well as maintenance and replacement problems, tend to set an upper limit, beyond which duplication of plant or of process units is preferable to expansion of a single unit or plant capacity. Similarly, for activities of the types listed in Table 3, there are minimum scales of output, below which the activity must generally be operated with different and less efficient methods. It is in the range between the minimum and maximum scales that the fractional "plant factor," as usually conceived, is relevant. Below the minimum scale, productive conditions result in a very steeply inclined unit cost curve for the activity.[8] Above the maximum scale, operation of a single plant runs into rising unit costs, frequently sharply rising because of the physical limitations. To preclude such rising costs, facilities are typically duplicated; in essence, at the maximum scale the plant factor becomes unity.[9]

In column 1 of Table 4 appear estimated capital-investment requirements for unit-level scales of operation of the activities listed at the extreme left of the table. (In terms of the fractional power law stated above, these investments may be taken as the I_k for the activities noted.) In column 2 are listed the plant factors for the activities.[10]

[8] Even in the scale range over which one plant factor applies, the corresponding unit-cost curve tends to be much more steeply inclined at lower scales than at higher. For this reason the minimum *economic* scale of an activity is often considerably greater than the minimum level to which the plant factor applies. (See p. 66.) Thus, in later chapters, instances are pointed out in which, at assumed final output levels, specific complexes require the operation of certain activities at smaller than minimum economic scale. Nevertheless, the capital investment costs for these activities are estimated by applying the same plant factors as are applicable in the output ranges of the activities above the economic minimum.

[9] For further discussion of the plant factor and associated concepts see Isard and Schooler, *op. cit.*, pp. 21–22; Lindsay, *op. cit.*, pp. 281–284.

[10] For any given activity only one plant factor is recorded. Unless specifically noted to the contrary, all subsequent calculations of capital investment for the activity pertain to output scales within the range covered by the recorded value of the plant factor.

Estimates of individual plant factors are based principally on Isard and Schooler, *op. cit.* especially Appendix B. Sources of information concerning investment-cost amounts are enumerated in Appendix C of the present study.

For four individual activities (44, 45, 63, 71) the estimated plant factors and investment amounts are derived from information in Airov, *op. cit.*

INDUSTRIAL COMPLEX ANALYSIS

TABLE 4
CAPITAL AND LABOR FACTORS AND UNIT LEVEL REQUIREMENTS

Activity	Capital Investment for Plant Operation at Unit Level (MM $)	Capital (plant) factor	Chem-petroleum labor required at unit level 100,000 mhr/yr.	Chem-petroleum labor factor	Textile labor required at unit level 100,000 mhr/yr.	Textile labor factor
	(1)	(2)	(3)	(4)	(5)	(6)
1. Oil ref. prototype #1	*	*	1.518	0.25	0	-
2. Oil ref. prototype #2	*	*	1.610	0.25	0	-
3. Oil ref. prototype #3	*	*	2.046	0.25	0	-
4. Oil ref. prototype #4	*	*	2.138	0.25	0	-
5. Oil ref. prototype #5	*	*	2.201	0.25	0	-
6. Oil ref. prototype #6	*	*	1.980	0.25	0	-
7. Ethylene separation (#1)	*	*		-	0	-
8. Ethylene separation (#2)	*	*	incl.	-	0	-
9. Ethylene separation (#3)	*	*	in oil	-	0	-
10. Ethylene separation (#4)	*	*	ref.	-	0	-
11. Ethylene separation (#5)	*	*	acti-	-	0	-
12. Ethylene separation (#6)	*	*	vity	-	0	-
13. Propane from LPG (dummy)	*	*	0	0	0	-
14. Ethylene from ethane	1.26	.67 ⎫	0.432	0.20	0	-
15. Ethylene from propane	1.39	.67 ⎭				
16. Chlorine by electrolysis	2.43	.75	0.655	0.218	0	-
17. Hydrochloric acid from Chlorine	0.78	.70	0.398	0.218	0	-
18. Hydrated lime	0.052	.70	0.916	0.218	0	-
19. Ethylene oxide (chlorhydrin)	2.05	.625	0.406	0.22	0	-
20. Ethylene oxide (oxidation)	2.56	.625	0.457	0.22	0	-
21. Ethylene Glycol (chlorhydrin)	2.06	.625	0.412	0.22	0	-
22. Ethylene Glycol (oxidation)	2.25	.625	0.427	0.22	0	-
23. Ethylene dichloride	2.77	.70	0.173	0.25	0	-
24. Acetylene from Methane	2.56	.60 ⎫				
25. Acetylene from ethylene	1.96	.60				
26. Acetylene from ethane	1.96	.60 ⎬	0.643	0.15	0	-
27. Acetylene from propylene	2.02	.60				
28. Acetylene from propane	2.02	.60 ⎭				
29. Vinyl chloride (ex ethylene di.)	1.48	.72	0.272	0.20	0	-
30. Vinyl chloride (ex acetylene)	1.48	.72	0.198	0.20	0	-
31. Ammonia from hydrogen	0.72	.80 ⎫				
32. Ammonia from methane	0.79	.80				
33. Ammonia from ethylene	0.79	.80				
34. Ammonia from ethane	0.79	.80 ⎬	0.260	0.40	0	-
35. Ammonia from propylene	0.79	.80				
36. Ammonia from propane	0.79	.80				
37. Ammonia from hy. resid.	0.82	.80				
38. Ammonia from cycle oil	0.82	.80 ⎭				
39. HCN from Methane and Ammonia	0.78	.70	0.135	0.20	0	-
40. Acrylonitrile from acetylene - HCN	0.92	.76	0.410	0.25	0	-
41. Acrylo. from ethylene oxide	0.69	.70	0.434	0.20	0	-
42. Sulfuric acid from sulfur	0.11	.85	0.082	0.218	0	-

INDIVIDUAL ACTIVITIES: INPUTS AND OUTPUTS

TABLE 4 (Cont'd)

	(1)	(2)	(3)	(4)	(5)	(6)
43. Nitric acid from ammonia	1.36	.63	0.123	0.20	0	–
44. Dimethylterephthalate (air ox.)	3.98	.67	0.875	0.218	0	–
45. Dimethylterephthalate (nitric acid ox)	3.98	.67	0.765	0.218	0	–
46. Dacron polymer	*	*	0.306	0.20	0	–
47. Dacron staple	*	*	1.450	0.60	8.250	1.00
48. Dacron filament	*	*	3.120	0.60	17.680	1.00
49. Dynel polymer	*	*	0.306	0.20	0	–
50. Dynel staple	*	*	1.450	0.60	8.250	1.00
51. Dynel filament	*	*	3.120	0.60	17.680	1.00
52. Orlon polymer	*	*	0.306	0.20	0	–
53. Orlon staple	*	*	1.450	0.60	8.250	1.00
54. Orlon filament	*	*	3.120	0.60	17.680	1.00
55. Ammon. nitrate from ammonia	0.13	.68	0.202	0.27	0	–
56. Urea from ammonia	0.72	.67	0.326	0.20	0	–
57. Benzene from refinery	*	*	0.218	0.218	0	–
58. Cyclohexane from benzene	*	*	0.164	0.218	0	–
59. Adipic acid from cyclohexane (nitric acid ox.)	*	*	0.926	0.218	0	–
60. Adipic acid from cyclo (air ox.)			0.764	0.218	0	–
61. Furfural from bagasse	*	*	0.655	0.218	0	–
62. Tetrahydrofuran from furfural	*	*	0.491	0.218	0	–
63. Sodium cyanide from HCN and Caustic	0.81	.65	0.436	0.218	0	–
64. Adiponitrile from Tetrahydrofuran	*	*	0.327	0.218	0	–
65. Butadiene from butane	*	*	1.900	0.218	0	–
66. Butadiene from refinery	*	*	2.240	0.218	0	–
67. Adiponitrile from butadiene	*	*	0.982	0.218	0	–
68. Adiponitrile from adipic acid			0.132	0.218	0	–
69. Adipic acid from Adiponitrile	*	*	0.545	0.218	0	–
70. Hexamethylene diamine from Adiponitrile	*	*	0.327	0.218	0	–
71. Nylon Salt	12.90	.60	0.153	0.218	0	–
72. Nylon staple	*	*	1.756	0.55	8.250	1.00
73. Nylon filament	*	*	3.426	0.50	17.680	1.00

*Estimate not required for the purposes of this study.

Sources: See text

Thus, the capital investment required for any activity at any level called for in complexes considered in this study can be obtained by multiplying the capital amount for the activity listed in column 1 by the required level of the activity raised to the power indicated by the plant factor. (As defined above, the required level of the activity is a multiple of the unit level, i.e., is given by the ratio O/O_k where O_k is taken as unit level.) For example, if ethylene dichloride is required at the rate of 40 MM lb/yr (i.e., at the level 4 since 10 MM lb constitutes unit level as defined in Table 3), then, since by Table 4 the relevant I_k and α are, respectively, $2.77 MM and 0.70, the estimated investment cost of the ethylene dichloride unit is

$$I = 2.77(4)^{0.70}$$

or $7.3 MM.

In general, the total capital investment for any set of activities (with output scales stipulated) can be calculated as the sum of the capital investments required by the individual activities, estimated as just described. However, there is one exception to this procedure. In the production of ethylene, acetylene, or ammonia, the different individual activities identify the various individual hydrocarbon feedstocks. But for each of these three commodities, the relevant individual feedstocks can be combined in one single productive process. Thus the capital investment requirements for ethylene, acetylene, or ammonia production must be calculated on the basis of the scale of this single productive process, which scale is equal to the sum of the scales of all the relevant individual activities.[11]

DIRECT LABOR INPUTS

As in the case of investment cost, industrial experience indicates that direct operating labor requirements for the activities considered in this study do not vary linearly with activity scale. The reasons are similar to those advanced above in the discussion of the variation of capital investment with scale. Doubling the capacity of a reactor does not double the man-hours of direct labor required for its operation. It is postulated in this study—as in most petrochemical studies—that direct labor inputs for the activities considered vary with scale in accordance with a fractional power law of the type

$$L = L_k \left(\frac{O}{O_k}\right)^\beta \qquad (0 < \beta \leq 1)$$

where L represents the labor requirements to be estimated for the

---

[11] For example, if a given program calls for the production annually of 15 MM lb of acetylene from methane, 15 MM lb from ethane, 10 MM lb from propane, and 5 MM lb each from ethylene and propylene, the total scale of the combined acetylene facility is 50 MM lb/yr. The total capital investment can be estimated by computing for each individual acetylene activity the investment required at a scale of 50 MM lb/yr, multiplying each resulting figure by the relevant percentage of total acetylene output accounted for by that activity, and summing the resulting amounts. Thus, the capital investment attributable to the "acetylene-from-methane" activity would be 30% of the investment required by a 50 MM lb/yr "acetylene-from-methane" operation, etc.

INDIVIDUAL ACTIVITIES: INPUTS AND OUTPUTS

projected plant of capacity O, and where L_k represents labor requirements for a known plant of scale O_k. Any given value of β the "labor factor," applies only over a certain range of output scales. Operation of the activity below or above this range would tend to involve a reorganization of the plant labor force, generally on a much less efficient basis.[12]

Skilled labor of various types is frequently in short supply in underdeveloped areas, while relatively unskilled labor tends to be plentiful. In the present study, therefore, two categories of labor are identified: textile labor and chemical–petroleum labor. It is assumed that Puerto Rico's supply of labor suitable for textile operations is more than adequate to meet the needs of any complex considered. On the other hand, it is assumed that all chemical-petroleum labor must be imported from the United States mainland. Such an assumption may appear to the reader as oversevere, but it is in line with the general intent of the study to be conservative in the estimation of Puerto Rico's locational advantages.

In columns 3 and 5, respectively, of Table 4 are listed for each activity considered the direct chemical–petroleum labor and the direct textile labor required for unit-level operation of the activity. Columns 4 and 6, respectively, record the chemical–petroleum labor factor and the textile labor factor for each activity.[13] Those factors that apply to synthetic fiber activities (staple and filament) —activities that require both chemical–petroleum and textile labor —are of particular interest. The inputs of textile labor in synthetic fiber activities are associated with a factor of unity. This linear

---

[12] Published data concerning labor inputs for industrial activities at different scales are scarce. One systematic analysis of the problem appears in H. E. Wessel, "New Graph Correlates Operating Labor Data for Chemical Processes," *Chemical Engineering,* **59,** 209-210 (July 1952). Wessel's graph is based on a fractional power law with an exponent of 0.218. Similar values for the labor factors can be calculated from a detailed analysis of three refineries of varying sizes made by the M. W. Kellogg Co., published in the *Kellogram,* 1954 series, issue no. 1. The empirical evidence of the existence of a fractional power law for capital inputs furnishes further justification for the acceptance of a formally analogous law for labor inputs. Obviously, for any given individual activity the value of the labor factor may not be in the neighborhood of 0.2 which is typical of many. In fact there is some evidence that the extreme values, 0 or 1, may characterize certain activities even in the relevant range of output scales.

[13] The estimated labor inputs and labor factors for the oil refinery activities (1–6, and 57) are based on data in Lindsay, *op. cit.* For synthetic fiber activities and for most of the chemical intermediate activities in nylon and Dacron production (44–54, 58–64 and 67–73), the labor inputs and factors are taken from Airov, *op. cit.* For the remainder of the activities, the recorded labor inputs and factors are based on information in Isard and Schooler, *op. cit.,* on other individual references cited in Appendix C, and on Wessel, *op. cit.*

relation of textile labor with scale reflects the fact that increases in activity above a very low limit of operation involve a duplication of identical process units. In contrast, inputs of chemical–petroleum labor are associated with a factor considerably less than unity. But, because of the duplication of process units at low levels of synthetic fiber operation, the chemical–petroleum labor factor of 0.6 which applies to synthetic fiber processes is at the same time substantially greater than the chemical–petroleum labor factor of 0.2 typical of most chemical processes.

The input of direct chemical–petroleum labor required by any activity at any level called for in complexes considered in this study can be calculated in accordance with the law stated above. For that activity, the relevant figure in column 3 of Table 4 is multiplied by the required level of the activity (as a multiple of the unit level) raised to the power indicated by the labor factor in column 4. If an activity requires direct textile labor, the calculation is exactly similar, except that the unit level input of column 5 and the labor factor of column 6 pertain.

In general, the total direct labor requirements of any set of activities is equal to the sum of the direct labor inputs required by the individual activities of the set calculated as described. However, as in the estimation of capital-investment requirements, an exception to the general procedure must be observed in calculating direct labor inputs for ethylene, acetylene, and ammonia production. The different activities listed for each of these three commodities refer to different possible feedstock sources. Since the various feedstocks for each commodity can be combined in a single productive process, the sum of the scale levels of the several individual activities, as noted before, is the relevant magnitude. At the same time, as depicted in Table 4, the same unit-level labor input and the same labor factor pertain to each individual activity of an ethylene, acetylene, or ammonia group. Thus the total direct labor input for any such group is calculated by applying the given labor factor to the combined activity scale and multiplying by the given unit-level labor input.

INDIRECT COSTS BASED ON CAPITAL AND DIRECT LABOR INPUTS

In addition to direct labor inputs and capital service inputs, activities of the type considered in this study require several process inputs whose costs are generally expressed as percentages of direct labor cost, capital investment cost, or both. Table 5 enumerates these indirect cost items. It records their typical estimated magnitude for chemical-type processes in terms of either direct labor or capital investment, or both.[14]

One way to treat these inputs for any activity would be to increase the amount recorded for total direct labor inputs and the amount

TABLE 5

ESTIMATES OF INDIRECT COSTS BASED ON DIRECT LABOR COSTS AND CAPITAL INVESTMENT (1)

ITEM	FRACTION OF L [2]	FRACTION OF C [2]
Supervision	0.1000	0.0000
Maintenance	0.0000	0.0400
Payroll Overhead [3]	0.1650	0.0030
Equipment and Operating Supplies	0.0000	0.0060
Indirect Production Cost [4]	0.5500	0.0200
General Office Overhead	0.1100	0.0046
Depreciation	0.0000	0.1000
Taxes	0.0000	0.0100
Insurance	0.0000	0.0100
TOTAL	0.925	0.1936

SOURCE: See text.

NOTES:

(1) To obtain cost of each item add the relevant fractions based on direct labor costs and based on capital investment.

(2) L = direct operating labor costs; C = investment in plant and equipment.

(3) Fringe benefits.

(4) First aid facilities, transportation in plant, safety, sanitation, analytical and technical services, maintenance of roads and yard, operation of general stockroom, utilities of nonoperating areas, non-specific maintenance expense.

---

[14] The figures recorded in Table 5 are based on Isard and Schooler, *op. cit.,* pp. 21–23. The principal references cited therein are R. S. Aries and associates, *Chemical Engineering Cost Estimation,* Chemonomics, New York, 1951; and U. S. Bureau of Mines, *Report of Investigations 4534,* "Guide for Making Cost Estimates for the Chemical Process Industries," Washington, 1945.

representing annual capital service inputs (i.e., total capital investment multiplied by the interest rate) by the factors indicated in the total row of Table 5. Thus the gross labor input would be the direct labor input multiplied by 1.925; the gross capital service input would be total capital investment multiplied by the sum of the interest rate plus 0.1936.

However, such a procedure is not entirely satisfactory from the standpoint of the present study. It is to be expected that a certain part of the costs based upon direct labor inputs will be associated with imported labor and a wage rate comparable to that of chemical-petroleum labor. The remaining part of the costs based upon direct labor inputs will be associated with local labor and a wage rate comparable to that of textile labor. Hence, it is necessary to break down the fractional increment applicable to each cost component into two subfractions: one referring to imported labor, and the other to local labor. Such a breakdown may differ between chemical-petroleum and textile activities. On the basis of the sparse information available, the breakdown of Table 6 is employed.

The increment of imported labor for an activity is added to the direct chemical-petroleum labor input for that activity since each is assumed to bear the same wage rate. The increment of local labor is added to the direct textile labor input for the same reason. Thus for any given activity (1) the gross imported labor input is made up of the direct chemical labor input multiplied by 1.470, plus the direct textile labor input multiplied by 0.305; (2) the gross local labor input is equal to the direct textile labor input multiplied by

TABLE 6

SELECTED INDIRECT COSTS: EXPRESSED IN TERMS OF DIRECT LABOR, IMPORTED AND LOCAL

	Fraction of Imported Labor	Fraction of Local Labor	Total
Supervision	0.050	0.050	0.100
Payroll overhead:			
for chemical labor inputs	0.165	0.000	0.165
for textile labor inputs	0.000	0.165	
Indirect Production cost	0.200	0.350	0.550
General office overhead	0.055	0.055	0.110
Total:			
for chemical labor inputs	0.470	0.455	0.925
for textile labor inputs	0.305	0.620	

INDIVIDUAL ACTIVITIES: INPUTS AND OUTPUTS 61

1.620, plus the direct chemical labor input multiplied by 0.455. It can be seen that every activity thus has both imported and local labor inputs, although for the majority of activities the direct operating labor input is solely of the chemical–petroleum type.

With respect to such cost items as depreciation, maintenance, and general office overhead, which are estimated partly on the basis of capital costs on investment (see Table 5), it must be recognized they will to some extent reflect transport charges incurred on plant and equipment units. In the case of Puerto Rico such charges could be significant, since many of these units would be shipped from the mainland. Unfortunately, the comprehensive and detailed material on physical input flows, etc. is not available, whereby such transport charges could be taken into account in the capital-cost analysis. Also, for the same reason, the analysis cannot take into account the fact that a significant portion of maintenance expense represents labor inputs, or the fact that the same may be true to some extent with regard to depreciation. As a result, the later cost analysis which assumes zero price differences on capital services can only be approximately accurate.

Chapter 4

The Full Programs

In Chapter 2 consideration of Puerto Rico's resources suggested the advantages of an industrial complex comprising oil refinery, petrochemical, and synthetic fiber activities. In Chapter 3 the technological data on physical inputs and outputs of the diverse individual production activities which may appear in such a complex were presented. The next step in the analysis, which is undertaken in this chapter, is the construction of the specific full programs or complexes—definite, feasible combinations of activities.

PROGRAM RESTRICTIONS

In view of the large number of activities listed in Table 3, and of the fact that most of these activities make up *alternatives* within one or more subgroups, it is obvious that the number of possible programs is very large indeed. This is particularly true if output scales of the diverse activities are subject to no limitation. However, it is possible by the use of certain empirical observations and reasonable economic judgments to narrow appreciably the set of possible specific programs that need to be investigated. Such observations and judgments form the basis for a number of restrictions observed in the construction of the programs considered in this study. Each of the three major activity groups—the refinery activities, the petrochemical activities (including fertilizer component production), and the synthetic fiber activities—is subject to definite constraints.

For reasons already indicated, the programs considered in this study are limited to those based on refinery prototype 4.[1] Furthermore, this prototype refinery is operated at unit scale only, i.e., at 9.428 MM bbl/yr crude oil throughout.[2] Such a restriction does not necessarily indicate the upper limit for a Puerto Rico refinery; rather it indicates the scale of a moderate-size, integrated refinery operation. Because such a refinery is economically feasible on the mainland, the comparison of its locational advantages on the mainland and in Puerto Rico is essentially realistic without further correction. Thus, the acceptance of the refinery scale limitation permits the study to be oriented toward the identification of the petrochemical possibilities associated with a single refinery neither overly large nor so small as to suffer serious diseconomies of scale.[3] The refinery scale restriction not only limits conveniently the total scope of the study, but also poses a regional comparison situation which has substantial empirical validity. The evaluation of a moderate-size refinery at a new location represents a cautious approach to a situation the risks of which can only partially be predicted. Such evaluation might well be the initial move of a large integrated United States firm considering progressive expansion of its operations in Puerto Rico. Alternatively, a similar evaluation might be undertaken by investigators interested principally in the probable course or pattern of Puerto Rico's economic development.[4]

Another constraint adopted in the study is that all final synthetic fiber activities have an output scale of 36.5 MM lb/yr. This scale is in a range of capacity common for mainland synthetic fiber plants. The reasoning behind the adoption of this scale constraint is essentially analogous to that described in the preceding paragraph with reference to the refinery scale. Thus, it can be said that this study is concerned with the evaluation of a number

---

[1] See pp. 49n–50n for explanation of this restriction.

[2] The daily crude oil throughput is 28,570 bbl. This specific rate is required in order to operate the catalytic cracking unit at a rate of 15,000 bbl/day.

[3] It is, of course, entirely possible, within the framework of this study, to modify or eliminate the refinery scale restriction and explore programs based on more than one refinery or on larger refineries.

[4] The restriction on refinery scales is raised by approximately 12% in the case of three individual programs (Orlon C, Orlon D, and Dynel E). Strictly speaking, all the other programs, to be entirely comparable, should be recalculated on the basis of the higher refinery scale. However, since the refinery does not contribute significantly to locational advantage, the slight raising of the refinery scale restriction has an insignificant effect upon the results of a comparison of the alternative programs.

of alternative petrochemical programs, linked at one end to a refinery of a given scale, and at the other end to synthetic fiber plants of a given scale.

Among the eight synthetic fiber production activities which appear in Table 3, four produce staple fiber and four produce continuous filament. In every case, the continuous-filament process is more advantageous for Puerto Rico, since it requires a much higher input of textile labor. Thus, whenever a complex containing a staple fiber activity possesses a net locational advantage, the same complex with the corresponding filament activity possesses an even greater net advantage. The detailed evaluation in succeeding chapters of individual complexes proceeds on the basis of staple fiber activities only. However, when the advantage of a staple fiber program is known, it is easy to determine the additional advantage that would accrue from the substitution of a filament activity for the staple fiber activity. The amount of this additional advantage is identical for Orlon, Dynel, Dacron, and nylon programs, since the additional requirements for textile labor are the same for all fibers when a filament substitutes for a staple fiber activity.[5]

Another constraint concerns the production of fertilizer components. The profitability of producing fertilizer in Puerto Rico would be sharply reduced if external markets, either in the Caribbean or on the mainland, were to be served. With respect to the fertilizer exported, Puerto Rico would suffer heavy transport-cost disadvantages and would probably be subject to certain other production-cost disadvantages compared to producers in the market region. Therefore, the production of fertilizer components in any complex of this study is required not to exceed the amount estimated as needed to meet local consumption. The estimate of such consumption is based upon data on Puerto Rico's fertilizer imports in 1952.[6]

Table 7 records the imports of various fertilizers and the amounts of ammonia that correspond to these imports on the basis of equivalent nitrogen content. These amounts total to 61.8 MM lb/yr. For purposes of this study, the local consumption estimate has been increased to 80 MM lb/yr in order to allow for some increase in fertilizer use. If the distant future is to be considered, a much greater local consumption estimate should be employed.

---

[5] The amount of the additional advantage is computed in Chapter 6.

[6] See "United States Trade in Merchandise and Gold and Silver with United States Territories and Possessions," *Census Report FT800,* 1953.

TABLE 7
PUERTO RICO FERTILIZER IMPORTS, 1952 AND CORRESPONDING QUANTITIES OF AMMONIA EQUIVALENT IN NITROGEN CONTENT

Commodity	Fertilizer Import (MM lb)	Ammonia Equivalent in Fixed Nitrogen Content	
		Percent of Fertilizer Weight	Estimated Quantities (MM lb)
Ammonium Sulfate	203.4	25.8 [a]	52.3
Nitrogenous fertilizer material, NEC	3.1	35.0 [b]	1.1
Nitrogenous phosphatic fertilizer material	82.2	10.0 [b]	8.2
Fertilizer mixtures, prepared	3.8	5.0 [b]	0.2
			61.8

[a] Calculated from chemical formula weights
[b] Estimated

The production of nitrogenous fertilizer is represented in this study by the ammonium nitrate and urea activities, which fit efficiently into a refinery–petrochemical setup. These two materials contain nitrogen in more concentrated form than does ammonium sulfate, the material which comprises the largest fraction of Puerto Rico's 1952 fertilizer imports.[7] Although ammonium nitrate and urea are considered as two separate production activities, their outputs in each individual program are scheduled in a one-to-one ratio, on a nitrogen-equivalent basis. This ratio was adopted in the absence of convincing reasons that would indicate a dominance of one or the other fertilizer form.

Another general restriction to which the various individual programs are subject is that the production of petrochemical intermediates (e.g., acetylene, acrylonitrile, ethylene, ammonia) is limited to the internal requirements of the program. Thus, each

---

[7] It is assumed that the nitrogen (or ammonia) equivalent is the dominant consideration in the evaluation of nitrogenous fertilizers, and that the different forms in which the nitrogen content may be embodied is of little consequence. It is possible, however, that such secondary properties of a nitrogenous fertilizer as acidity–alkalinity or ease of application may be significant to its agricultural usefulness. Furthermore, the persistence of traditional methods may cause a preference for ammonium sulfate. Nevertheless, the influence of all such factors is conjectural, and for that reason they have been ignored.

program is designed to produce a zero (or at most a negligible) amount of these commodities as outputs available for sale. The reason for this restriction is that, initially at least, a Puerto Rico complex would not encounter significant local demand for petrochemicals external to the complex itself. With respect to mainland markets, Puerto Rico complexes would suffer substantial transport- and production-cost disadvantages, in addition to the difficulties involved in penetrating the mainland market for petrochemicals.[8]

Because of their technological characteristics, virtually all the individual activities of refinery–petrochemical–synthetic fiber complexes must, generally speaking, be operated at or above some minimum scale in order to be economically feasible. This minimum scale usually falls somewhere in the output range within which unit costs begin to rise sharply with decreases in scale. The minimum economic scale does not necessarily coincide with the scale below which a given plant or labor factor ceases to be relevant. At small output scales, unit costs may vary sharply with scale, even though given plant and labor factors still apply. Thus, an important restriction to which the programs considered in this study are subject is that, aside from certain special cases, no program may contain any activity scheduled at a smaller scale than its minimum scale which is recorded in Table 8.[9] (The minimum scales of Table 8 are expressed as multiples of the unit level outputs which appear in Table 3.)[10]

---

[8] In a few cases, however, chemical by-products resulting from the operation of the petrochemical activities are considered to be exported to the mainland or sold locally. See Chapter 5.

[9] In some of the programs which have been calculated in detail, minimum-scale constraints for certain activities have been violated in order to obtain a group of programs representative of the major possible process chains. These violations are noted and discussed in later chapters.

[10] The minimum-scale estimates are based on a variety of sources including published information, estimates of chemical engineers, and correspondence with chemical firms. Limits for activities 7–41, 43, 45, 56 are from Walter Isard and Eugene W. Schooler, *Location Factors in the Petrochemical Industry,* U. S. Department of Commerce, Office of Technical Services, 1955, pp. 22–24 and Appendix B. The lower refinery limit (activities 1–6) represents the decision to program only unit scale (i.e., moderate-size) refineries.

Limits for activities 46–54 and 71–73 are from Joseph Airov, *The Location of the Synthetic Fiber Industry,* John Wiley & Sons, 1959. Limits for activity 42 are from W. L. Faith, D. B. Keyes, and R. L. Clark, *Industrial Chemicals,* 2nd ed., John Wiley & Sons, 1957, p. 750. The limit is applicable to the contact process; this process yields concentrated acid as required, while the chamber process yields dilute acid. The limit for the chamber process is 20 MM lb/yr.

THE FULL PROGRAMS

For several of the petrochemical commodities of Table 3, alternative productive processes, listed as separate activities, are possible. This situation leads to the adoption of another constraint closely related to the type just discussed. The scheduling of more than one alternative activity for a given commodity in a given program is precluded. Such a constraint is based on the fact that economies of scale are gained when one process for the production of a given commodity is employed at a large scale rather than a group of two or more processes, each at a small scale. The relevant groups of activities are: 19 and 20, ethylene oxide via the chlorhydrin process and via the oxidation process; 29 and 30, vinyl chloride from ethylene and from acetylene; 40 and 41, acrylonitrile from acetylene and from ethylene oxide; 44 and 45, dimethyl terephthalate via the air oxidation process and via the nitric acid oxidation process; 59, 60, and 69, adipic acid from adiponitrile, from cyclohexane via air oxidation and from cyclohexane via nitric acid oxidation; and 64, 67, and 68, adiponitrile from tetrahydrofuran, butadiene, and acetic acid. It should be emphasized that this general type of constraint is not relevant for

---

No limit was estimated for activities 13, 16, 17, 18, 44, and 45, since the scales at which these activities are programmed were assumed to be sufficiently high so that a limit would not act as a constraint.

The limiting section of a combined ethylene process, embodying separation (7-17) and cracking (13, 14) is the separation plant. Even though activities 13 and 14 produce ethylene from other gases, the cracking reactor yields a mixed gas stream from which ethylene is separated, and part of the gases are recirculated. This separation facility can be combined with the separation facilities of activities 7-12; the lower limit for this separation facility is estimated by Isard and Schooler to be 25 MM lb/yr. It is assumed that the cracking furnace itself can be of considerably smaller scale; thus, for example, a program embodying 15 MM lb/yr of ethylene by, say, activity 10, and 10 MM lb/yr by activity 14, is acceptable. No lower limit on the cracking part of activities 14 and 15 is explicitly assumed, but about 5 MM lb/yr output would probably be a sensible lower limit to observe in actual programming. None of the programs developed in detail violates this limit.

The limit for activity 57 corresponds to approximately 500 bbl per day of benzene production. This is generally considered to be the lower limit under U. S. conditions. See the references cited on benzene production, Appendix A.

For activities 58-64 and 67-70, no detailed information was available on lower limits. These limits have arbitrarily been set at an output of 10 MM lb/yr of product. This corresponds to the limit of several other organic intermediates in Table 8.

Limits for activities 65 and 66 are taken from Faith, Keyes, and Clark, *op. cit.*, p. 178. They are based on 15,000 tons/yr production plus about 40%, as given in this reference.

TABLE 8
MINIMUM SCALES FOR ACTIVITIES: IN MULTIPLES OF UNIT LEVEL OUTPUTS

ACTIVITY	MULTIPLE	ACTIVITY	MULTIPLE
# 1 Refinery Prototype #1	1.0	# 37 Ammonia ex Heavy Residual ⎫	
2 Refinery " #2	1.0	38 Ammonia ex Cycle Oil ⎭	
3 Refinery " #3	1.0	39 Hydrogen Cyanide	1.0
4 Refinery " #4	1.0	40 Acrylonitrile ex Acetylene	0.5
5 Refinery " #5	1.0	41 Acrylonitrile ex Ethylene	0.5
6 Refinery " #6	1.0	42 Sulfuric Acid	1.7
7 Ethylene by Separation ⎫		43 Nitric Acid	1.0
8 Ethylene by Separation		44 Dimethyl Terephth: Ox.	NR
9 Ethylene by Separation	∗	45 Dimethyl Terephth: Nit. Ac.	NR
10 Ethylene by Separation		46 Dacron Polymer	3.0
11 Ethylene by Separation		47 Dacron Staple	3.0
12 Ethylene by Separation ⎭		48 Dacron Filament	1.0
13 Propane Dummy	NR	49 Dynel Polymer	3.0
14 Ethylene ex Ethane ⎫	∗	50 Dynel Staple	3.0
15 Ethylene ex Propane ⎭		51 Dynel Filament	1.0
16 Chlorine	0.365	52 Orlon Polymer	3.0
17 Hydrochloric Acid	NR	53 Orlon Staple	3.0
18 Lime	NR	54 Orlon Filament	1.0
19 Ethylene Ox. via Chlor.	1.0	55 Ammonium Nitrate	5.0
20 Ethylene Ox. via Ox.	1.0	56 Urea	3.0
21 Ethylene Glyc. via Chl.	1.0	57 Benzene from Refinery	2.0
22 Ethylene Glyc. via Ox.	1.0	58 Cyclohexane ex Benzene	1.0
23 Ethylene Dichloride	2.0	59 Adipic Ac. ex Cyhex. (nit.)	1.0
24 Acetylene ex Methane		60 Adipic Ac. ex Cyhex. (air)	1.0
25 Acetylene ex Ethylene		61 Furfural ex Bagasse	1.0
26 Acetylene ex Ethane	1.0	62 Tetrahydrofuran ex Furfural	1.0
27 Acetylene ex Propylene		63 Sodium Cyanide ex HCN	1.0
28 Acetylene ex Propane		64 Adiponitrile ex Tet.h.f.	1.0
29 Vinyl Chl. ex Ethylene	2.0	65 Butadiene ex Butane	4.0
30 Vinyl Chl. ex Acetylene	2.0	66 Butadiene ex Refinery	4.0
31 Ammonia ex Hydrogen ⎫		67 Adiponitrile ex Butadiene	1.0
32 Ammonia ex Methane		68 Adiponitrile ex Adipic Acid	1.0
33 Ammonia ex Ethylene	6.6	69 Adipic Ac. ex Adiponitrile	1.0
34 Ammonia ex Ethane		70 Hexamethylene Diamine	1.0
35 Ammonia ex Propylene		71 Nylon Salt	3.0
36 Ammonia ex Propane ⎭		72 Nylon Staple	3.0
		73 Nylon Filament	3.0

NR: No Restriction

∗ Total ethylene by separation (#7-#12) and cracking (#14,#15): 25MM lb/yr.

the ethylene, acetylene, or ammonia activity groups. The activities of each of these groups represent the use of alternative feedstock materials, all or any of which may be combined without loss of economy in the actual production process.

THE GENERAL PROGRAM PATTERN

The choice of individual programs for detailed evaluation in this study reflects two important objectives: (1) The programs must be feasible; i.e., they must satisfy the various constraints; (2) the programs should be representative of the major processing chains which lead from raw materials to finished products.

A program typical of those selected begins with the import of crude oil from Venezuela. Subsequently, there are four processing stages. From the last stage, the final products—synthetic fiber and fertilizer—emerge in finished form and are sold. The first stage consists of the petroleum refinery, which utilizes crude oil, labor, steam, fuel, and power inputs. The three utilities inputs are generated by burning part of the refinery products. (In fact, the required inputs of utilities for the whole complex derive from this source.) The remaining refinery output moves along two paths: one to market for sales, the other to petrochemical production activities.

The second major process is primary petrochemical production. The raw-material inputs to this stage come directly from the refinery, and some of the outputs are returned to the refinery. Thus, the interdependence between activities of this stage and of the refinery is strong. The main outputs, aside from by-products, are the six basic petrochemicals of the present study: ethylene, acetylene, methane, ammonia, benzene, and butadiene.[11]

The third principal processing stage is secondary petrochemical production. The six raw materials flowing into this stage are converted, often by a chain of steps, into various chemical intermediates, which in their turn serve as inputs for the final stage. For example, one of the routes leading to acrylonitrile (a raw material for Dynel and Orlon fiber) is

```
Ethylene    ⟶ Ethylene oxide    ⎫
Methane  ⎫                        ⎬ ⟶ Acrylonitrile
         ⎬  ⟶ Hydrogen cyanide  ⎭
Ammonia  ⎭
```

The final processing stage leads to finished products ready for sale. The finished products are: Orlon, Dynel, Dacron, and nylon

---

[11] The production of some equally basic petrochemicals, such as methyl, ethyl, and isopropyl alcohols, ethylbenzene, acetone, is not included among the activities of this study, since such commodities do not serve as intermediates for the selected fibers or fertilizers. Because of this exclusion, the procedure is not strictly correct. For example, when ammonia is produced, it is practical to consider the simultaneous production of methanol, since the processes have many common features and can use combined facilities. However, such considerations are of minor significance within the framework of this study. Also, omission of such considerations made possible a considerable narrowing of the field of inquiry.

fiber, and the nitrogenous fertilizers: ammonium nitrate and urea. The fibers are produced in two steps: first, the formation of a polymer (long-chain molecule); second, the extrusion, fixation, and finishing of the fiber.

The above four-stage outline fails to account for certain raw-material and intermediate chemical inputs. Limestone, salt, sulfur, and bagasse are treated as basic inputs which must be purchased outside the complex. Limestone and salt are assumed to be ubiquities which are available at an equal price in Puerto Rico and on the mainland; sulfur is assumed to be imported to Puerto Rico at a price equal to the mainland price plus transport cost. Bagasse is assumed to be available in Puerto Rico as a by-product of sugar production. Carbon dioxide, while essentially a chemical intermediate, is formally treated as a basic input available at equal cost in both locations.[12] In addition to these, there are three chemicals, acetone, dimethyl formamide, and paraxylene, that could conceivably be produced in Puerto Rico. However, activities producing these chemicals are not included in the analysis since the scale at which each would need to be undertaken in any feasible program (under the constraints discussed earlier) is well below the minimum economical scale.[13]

The four-stage programming scheme yields programs that proceed by a sequence of steps from the refinery to the fiber and fertilizer processes. However, alternative programs can be created

[12] See Chapter 5.
[13] Acetone and dimethyl formamide occur as inputs of Dynel and Orlon manufacture. The inputs of these chemicals for 36.5 MM lb/yr fiber production are 10.95 and 8.40 MM lb/yr, respectively (see flowsheets and Table 11). A scale of 10.95 MM lb/yr is clearly much too small for acetone, which is a heavy industrial chemical. A scale of 8.40 MM lb/yr for dimethyl formamide also appears to be insufficient, although the minimum-scale limit for dimethyl formamide cannot be too clearly established.

Regarding paraxylene, the following estimates show the order of magnitude involved. A unit-scale refinery (charging 28,570 bbl/day crude) will yield, by naphtha re-forming and Udex extraction processes, a maximum of about 25 MM lb/yr of mixed xylenes. The paraxylene yield on mixed xylenes is about 10 to 15%; but, even counting the double of this yield, the available paraxylene (25 MM)(0.3) = 7.5 MM lb/yr, falls far short of the necessary 21–25 MM lb/yr required for the Dacron programs. At the same time, the scale of operations would be too small. Consequently, paraxylene is considered to be imported for all programs. The sources of data on xylenes are: (1) Davis Read, "The Production of High Purity Aromatics for Chemicals," *Petroleum Refiner,* **31,** 97–103 (May 1952); (2) V. B. Guthrie, "Man-Size Petrochemical Venture," *Petroleum Processing,* **9,** 83 (1954); (3) V. K. Jackson et al., "Making High Purity Aromatics," *Petroleum Processing,* **9,** 233 (1954); (4) private communications from petroleum companies.

by substituting imports for production anywhere along the chain. In fact, the set of feasible programs includes programs consisting of only one, two, or a few activities. Hence, to cover the entire set would require programming for all conceivable combinations of activities. However, the authors have chosen, for reasons already stated, to evaluate first and most comprehensively the more completely integrated combinations, and later to examine in less detail combinations involving the substitution of imports for local production and comprising fewer activities.

DETAILED PRESENTATION OF SAMPLE PROGRAMS

In this section, a few programs are worked out in detail. All these programs embody the four principal stages discussed above; i.e., they comprise petroleum refining, primary petrochemical operations, secondary petrochemical operations, and the production of finished fiber and fertilizer. In these programs, no intermediates are imported for which a local production activity is given in Table 3 except for cyclohexane and butadiene in the production of nylon. These exceptions arise because the required scales for the production of cyclohexane and butadiene would be well below the minima recorded in Table 8. The programs are all based on the same refinery, activity 4 of Table 3.

Flowsheets and Subprograms

The programs are built around flowsheets 1, 2, 3, and 4, which directly follow. These flowsheets show the various routes leading from the six basic petrochemicals (ethylene, acetylene, methane, ammonia, benzene, and butadiene) to finished Orlon, Dynel, Dacron, and nylon fibers. The scale of the fiber activity corresponds in every case to 36.5 MM lb/yr fiber output, which is the maximum permissible under the fiber activity scale constraint. Programs with a smaller fiber output are not worth considering, since the fiber activities have a high input of textile-grade labor, and are very advantageous in Puerto Rico. If the fiber activities were scheduled at less than the 36.5 MM lb/yr limit, any over-all advantage which Puerto Rico may have would be understated.

The flowsheets indicate the amounts of intermediate commodities required at each step. For example, flowsheet 1 on Orlon indicates at the top that 36.5 MM lb of Orlon staple requires 8.40

72   INDUSTRIAL COMPLEX ANALYSIS

Flowsheet 1. Orlon

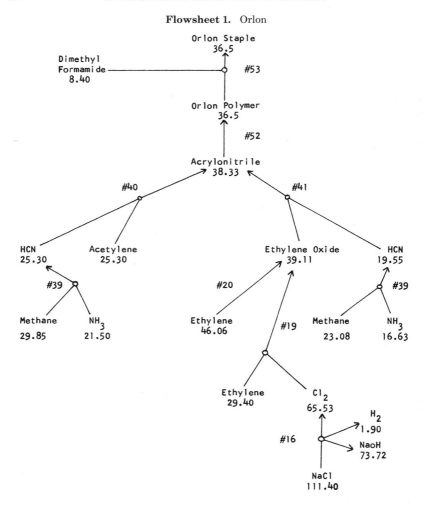

MM lb of dimethyl formamide (which is imported), and 36.5 MM lb of Orlon polymer (which is produced locally).[14] Note that, when more than one input are needed for an output, the lines originating at the inputs are pulled together in a bead, and a single arrow is drawn from the bead to the output commodity. Note also that the activity number which indicates the specific process using these inputs and which yields the relevant output, in this case 53, is placed by the appropriate arrow. The 36.5 MM lb of

---

[14] These data are based upon material contained in column 53 of Table 3.

THE FULL PROGRAMS 73

Flowsheet 2. Dynel

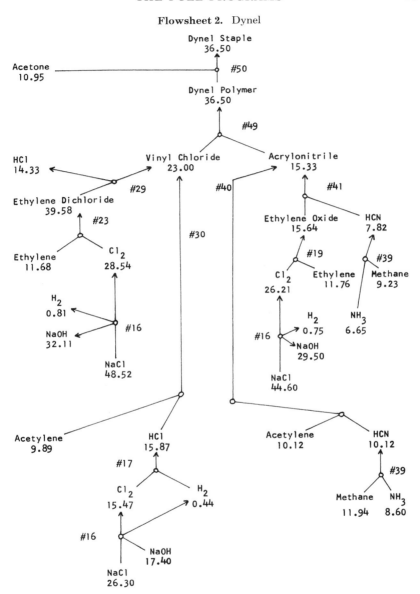

Orlon polymer requires, via activity 52, 38.33 MM lb of acrylonitrile, which in turn requires a set of inputs. Since acrylonitrile can be produced by either of two major ways (processes 40 and 41) as indicated by the two arrows leading to acrylonitrile in the flowsheet,

Flowsheet 3. Dacron

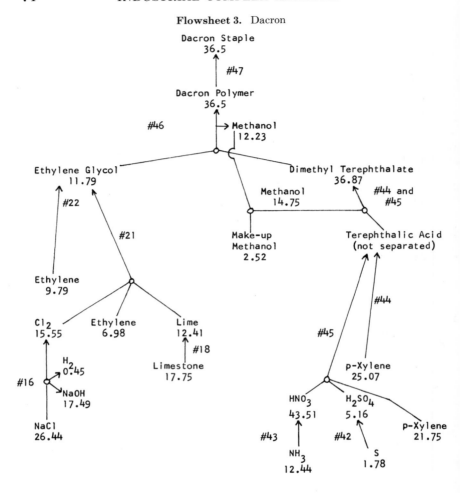

these inputs depend on the specific way chosen. In turn, these derived inputs of petrochemical intermediates require other intermediates; etc. as detailed in the flowsheet.

From the flowsheets, various alternative subprograms connecting the basic petrochemicals with the final fibers can be constructed. For example, from flowsheet 1 on Orlon can be constructed the Orlon subprograms noted at the bottom of each section in Table 9. (Subprograms on other fibers appear as Tables B–1 to B–3 in Appendix B.) The data for each subprogram of Table 9 are directly obtained from flowsheet 1. These data are reformulated in terms

THE FULL PROGRAMS

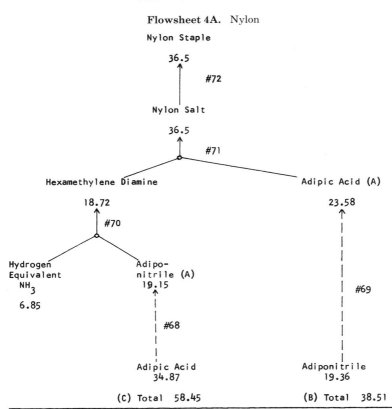

Flowsheet 4A. Nylon

Note: There are three main routes to Nylon. Route A: mixed base where activities #68 and #69 are absent and the adiponitrile and adipic acid requirements are obtained according to Flowsheets 4B and 4C respectively. Route B: all adiponitrile base where activity #69 is added in order to produce the required adipic acid and where the adiponitrile is produced according to Flowsheet 4B. Route C: all adipic acid base where activity #68 is added (activity #69 being absent) in order to produce the required adiponitrile and where the adipic acid is produced according to Flowsheet 4C.

of the unit levels (scales) of activities of Table 3. Thus the first subprogram (relevant for Orlon A, Orlon B, Orlon C, and Orlon D), which requires the production of

36.5 MM lb of Orlon staple,
36.5 MM lb of Orlon polymer,
38.33 MM lb of acrylonitrile,
39.11 MM lb of ethylene oxide, and
19.55 MM lb of hydrogen cyanide,

76 INDUSTRIAL COMPLEX ANALYSIS

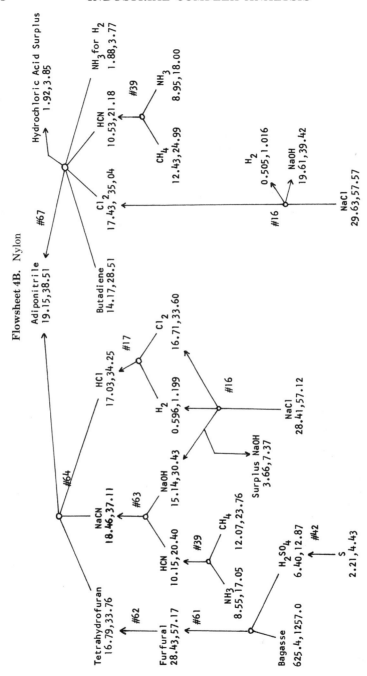

Flowsheet 4B. Nylon

Note: For each material, two numbers are given. The first of these refers to Route A: mixed base. The second refers to Route B: all adiponitrile base, when adipic acid is made from adiponitrile. See also Flowsheet 4A.

THE FULL PROGRAMS

Flowsheet 4C. Nylon

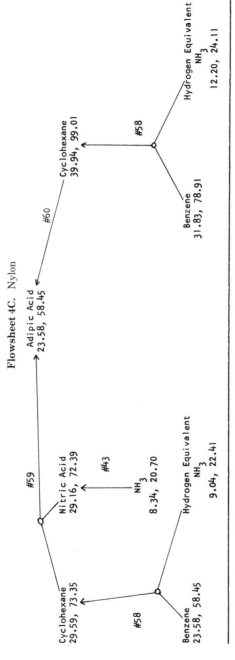

Note: For each material, two numbers are given. The first of these refers to Route A: mixed base. The second refers to Route C: all adipic acid base, when adiponitrile is made from adipic acid. See also Flowsheet 4A.

TABLE 9
ORLON SUBPROGRAMS

FOR ORLON A, B, C, D

Ethylene 46.06 → #20 Ethylene Oxide 39.11 ⎫
Methane 23.08 ⎫   ⎬ → #41 Acrylonitrile 38.33 → #52 Orlon Polymer 36.5
Ammonia 16.63 ⎭ → #39 Hydrogen Cyanide 19.55 ⎭   → #53 Orlon Staple 36.5

Subprogram: (3.65) (#52) + (3.65) (#53) + (3.833) (#41) + (3.911) (#20) + (1.955) (#39)

FOR ORLON E, F

Ethylene 29.40 ⎫
Chlorine 65.53 ⎭ → #19 Ethylene Oxide 39.11 ⎫
                                            ⎬ → #41 Acrylonitrile 38.33 → #52 Orlon Polymer 36.5
Methane 23.08 ⎫   ⎭   → #53 Orlon Staple 36.5
Ammonia 16.63 ⎭ → #39 HCN 19.55

Subprogram: (3.65) (#52) + (3.65) (#53) + (3.833) (#41) + (3.911) (#19) + (1.955) (#39)

FOR ORLON H, J

Acetylene 25.30 ⎫
Methane 29.85 ⎫ ⎬ → #39 HCN 25.30 → #40 Acrylonitrile 38.33 → #52 Orlon Polymer 36.5
Ammonia 21.50 ⎭   → #53 Orlon Staple 36.5

Subprogram: (3.65) (#52) + (3.65) (#53) + (3.833) (#40) + (2.530) (#39)

indicates that activity 52 (which yields at unit level 10 MM lb of Orlon polymer) must be multiplied by 3.65; activity 53 (which yields at unit level 10 MM lb of Orlon staple) by 3.65; activity 41 (which yields at unit level 10 MM lb of acrylonitrile) by 3.833; etc.[15]

---

[15] Note that the unit levels of most of the petrochemical activities are designed to yield 10 MM lb of output, so that for the most part only the shifting of decimal points is involved in detailing the appropriate subprograms.

Production Programs

Once the subprograms have been constructed, they must be supplemented in two ways: (1) The basic petrochemicals must be connected to the refinery;[16] (2) fertilizer activities and a few other productive activities, such as chlorine, hydrochloric acid, and lime production must be added. This supplementation is done according to a predetermined procedure, since the study is organized in such a way that the subprograms embody the main programming decisions.[17] Thus, the basic petrochemical requirements are connected with refinery 4 in a standard sequence, and fertilizer activities are added at a fixed level. The detail resulting from this procedure for each of the integrated complexes considered in this study are presented in Tables B-4 to B-28 in Appendix B.

In the construction of the production programs for Orlon, Dynel, Dacron, and nylon fibers, a number of steps were taken and decisions made which require further discussion.[18] As already noted, the subprograms (together, in the case of nylon, with a few additional steps in the production programs) specify the input requirements of the six basic petrochemicals: ethylene, acetylene, methane, ammonia, benzene, and butadiene. The last two occur only in nylon programs and are treated somewhat differently from the first four. Ethylene, acetylene, methane, and ammonia are typically produced from refinery output streams which contain a mixture of components. In the present study a highly simplified representation of refinery output has been adopted which reduces to the following the number of possible mixture streams going to the production of ethylene, methane, acetylene, and ammonia:

1. Cycle oil and heavy residual (commodities 10 and 11 of Table 3). These may go to ammonia production (activities 38 and 37).

2. Liquefied petroleum gas (LPG, commodity 13). This material may go to ethylene, acetylene, or ammonia production (activities 15, 28, 36) by way of a dummy activity (13). The purpose of this dummy activity is to reserve for possible petrochemical use some

---

[16] In the case of nylon, the subprograms are not carried all the way back to the basic petrochemicals; therefore the missing steps must be supplied subsequently.

[17] In the case of nylon, some major programming decisions are deferred to the production programs. The reasons are mainly expositional. Nylon has a more complicated flowsheet than the other fibers; complete subprograms would, therefore, be difficult to present compactly.

[18] The nontechnical reader may elect to skip the remaining discussion of this section.

of the propane which is usually fed into the LPG pool consisting primarily of propane and butanes.

3. A mixed gas stream, PLAT, which comes from the platforming section of the refinery. This stream is rich in hydrogen and also contains some methane and ethane. It is available for ammonia manufacture (activities 31, 32, 34).

4. A mixed gas stream, H_2C_1, which contains the light gases from all sections of the refinery other than platforming. Besides hydrogen and a very high proportion of methane, the stream contains some ethylene and ethane. It is available for hydrogen cyanide, acetylene, and ammonia manufacture (activities 39, 24-26, 31-34).

5. A mixed gas stream, C_2—C_2C_3—C_3, which contains ethylene, ethane, propylene, and propane. This stream originates in various sections of the refinery and is stripped of the lightest gaseous components (these are in the H_2C_1 stream) and also of a good part of the heavier components (these are in the LPG stream). The C_2—C_2C_3—C_3 stream is available for ethylene, acetylene, and ammonia manufacture (activities 7-12, 25-28, 33-36).

In Table 3, the last three streams are not represented by distinctly defined commodities, since the composition of these streams is different for each of the six refinery prototypes. Rather, the components of the streams are listed as commodities. Activities are defined in terms of the individual components as inputs; thus, for example, ammonia activities are defined separately for hydrogen, methane, ethylene, . . . ; etc. *The assumption is made that the activities defined for the individual components are additive when the components are present in a mixture.* Such assumption neglects the fact that the best reaction conditions for the various components are likely to be different from one another, and that a mixture will usually give poorer product yields than the components individually. However, allowance for this fact would require a refinement of technical data which was not possible in this study.

The activities involving the components of a stream are performed in proportion to the components available in the stream. Thus, when a H_2C_1 stream is sent to ammonia manufacture, the ratios of activities 31, 32, 33, and 34 are not freely variable, but are determined by the composition of the stream. Essentially, this requirement constitutes a specific constraint involving mixture ratios. It may conveniently be referred to as a "proportionality" constraint.

The compositions of the gaseous streams obtained from refinery 4 are given in detail in Table 10, since all the programs of the

TABLE 10
REFINERY GAS FROM REFINERY PROTOTYPE #4 (UNIT LEVEL OPERATION): HYDROCARBON COMPOSITION AND ALTERNATIVE USES

	TOTAL	COMPOSITION			AMMONIA FROM			ACETYLENE FROM		
		PLAT	H_2C_1	$C_2-C_2C_3-C_3$	PLAT	H_2C_1	$C_2-C_2C_3-C_3$	PLAT	H_2C_1	$C_2-C_2C_3-C_3$
Hydrogen	8.98	7.19	1.79	0	35.95	8.45	0	0	0	0
Methane	34.86	3.75	31.11	0	6.82	56.56	0	1.10	9.10	0
Ethylene	17.41	0	1.31	16.10	0	2.08	25.60	0	0.72	8.85
Ethane	32.25	0.66	1.40	30.19	1.14	2.42	52.23	0.34	0.72	15.48
Propylene	7.58	0	0	7.58	0	0	11.81	0	0	3.21
Propane	5.08	0	0	5.08	0	0	8.40	0	0	2.06
					43.91	70.01	98.04	1.44	10.54	29.60

study are based on this refinery. Furthermore, the amounts of ammonia and acetylene obtainable from the streams of a unit scale 4 refinery are included in this tabulation to aid the understanding of the structure of the production programs.

The production of ethylene from the C_2—C_2C_3—C_3 stream has been represented by a separate activity for each refinery (7–12). These activities produce pure ethylene by a process of separation from the mixed gas stream. A joint product is pure ethane.[19] Other activities for ethylene production are the cracking of pure propane (secured from the LPG output of the refinery by way of a dummy activity, 13) and the cracking of pure ethane, the by-product of ethylene separation.

A final stream which needs to be considered jointly with the foregoing ones, even though it originates outside the refinery, is electrolytic hydrogen, a by-product of chlorine manufacture. As petrochemical hydrogen this hydrogen may be processed into ammonia; in fact, it is much preferred because of its high purity. Furthermore, it can be used directly for the chemical hydrogenation steps in activities 58, 67, and 70, whereas the mixed hydrogen in the refinery gases cannot be so used without extensive separation and purification processes. Accordingly, the procedure followed in connection with the requirement for pure hydrogen is the following. First, the total amount of pure hydrogen needed is determined. Next, the available by-product electrolytic hydrogen is deducted. Finally, the remaining requirement of pure hydrogen is replaced by its ammonia equivalent on the assumption that the ammonia process is robbed of the small amount of pure hydrogen so needed.

The basic petrochemicals butadiene and benzene can be connected to the refinery by way of activities 57, 65 and 66. These activities are, for practical purposes, independent of the connections of the first four basic petrochemicals with the refinery.

Butadiene and benzene occur only in nylon programs. In the case of butadiene, the necessary scale of the connecting activities would be very low; hence in the programs actually calculated it is assumed that butadiene is imported. The production of benzene from refinery streams is assumed in some of the programs; in others, benzene production is replaced by cyclohexane imports. Actually, benzene imports could also have been assumed, but benzene, when it occurs, is converted to cyclohexane in every instance, and the import of cyclohexane is more advantageous than the im-

---

[19] Polymer gasoline and LPG are other joint products. For a discussion of the relationship of ethylene separation to the refinery, see Appendix A, pp. 225–227.

THE FULL PROGRAMS

port of benzene.[20] The scale of benzene production in some of the programs violates minimum-scale restrictions. This subject is discussed further in Chapter 6.

The procedure for obtaining the production programs detailed in Appendix B begins with the ancillary inorganic activities which are not included in the subprograms.[21] These activities include the production of hydrochloric acid (17), chlorine (16), lime (18), sulfuric acid (42), and nitric acid (43). These activities are programmed at a level sufficient to yield the required inputs for the subprograms.

Next, the basic petrochemicals are connected to the refinery. The scale of the refinery is chosen as unity. In three programs (Orlon C, Orlon D, and Dynel E), the refinery scale is chosen slightly higher in order to permit the production of all ethylene by separation (Dynel E), or to eliminate the need for propane cracking for ethylene production (Orlon C and Orlon D).

The connecting of benzene and butadiene to the refinery presents no difficulties and proceeds as discussed above.[22] Ethylene, methane, acetylene, and ammonia are independent of benzene and butadiene, and are connected to the refinery in a definite sequence based on the following tabulation.

By-Product Hydrogen	Plat	H_2C_1	C_2—C_2C_3—C_3
Ammonia(1)	Ammonia(2)	Methane Acetylene(3) Ammonia(3)	Ethylene(1) Acetylene(2) Ammonia(5)
By-Product Pure Ethane	LPG Propane		Fuel Oil
Ethylene(2) Acetylene(1) Ammonia(4)	Ethylene(3) Acetylene(4)		Ammonia(6)

[20] The benzene-to-cyclohexane production step has fuel and labor inputs. Thus a Puerto Rico location for this step would be at a disadvantage compared to a mainland location.

[21] In the case of nylon, the procedure begins with the steps leading from the basic petrochemicals to adiponitrile or adipic acid. These steps have been omitted from the subprograms for simplicity of exposition. Thereafter, the procedure is the same. In a few nylon programs, cyclohexane takes the place of benzene as the basic petrochemical on which further steps are based. In these cases, the cyclohexane is imported and not based on the refinery.

[22] As already indicated, these connections are replaced by imports in some programs.

This tabulation indicates, first, the hierarchy of claims by various products upon a given stream, and, second, the order of preference in which streams are utilized for a given product. The hierarchy of claims upon a stream is denoted by sequence within a column, the order of preference by numbers.

The above tabulation is based upon, among others, the following considerations relating to optimal programming: (a) Methane for hydrogen cyanide can be based only on a stream rich in methane, not on other streams; therefore, it has first claim on the H_2C_1 stream. (b) Ethylene production by separation has first claim on the C_2—C_2C_3—C_3 stream, since such separation is a more efficient use than cracking to acetylene; moreover, ethylene separation yields a by-product ethane stream which may be cracked to acetylene. (c) Ethylene production by cracking has first claim on the by-product ethane stream since acetylene unlike ethylene can be alternately produced from the H_2C_1 stream without cutting into the valuable LPG pool. (d) Ethylene production by cracking has first claim on propane since acetylene unlike ethylene can be produced from the H_2C_1 stream. (e) The order of preference for ethylene production is (1) C_2—C_2C_3—C_3 by separation, (2) by-product ethane, (3) LPG propane, because the first of these needs no cracking, while the last cuts into the valuable LPG pool. (f) The order of preference for acetylene is (1) by-product ethane, (2) C_2—C_2C_3—C_3 stream, (3) H_2C_1 stream, (4) LPG propane, because the first three are of decreasing technical efficiency, while the last cuts into the valuable LPG pool. (g) The order of preference for ammonia is (1) by-product hydrogen, (2) platformer light gas, (3) H_2C_1 stream, (4) by-product pure ethane, (5) C_2—C_2C_3—C_3 stream, (6) cycle oil, because this order indicates decreasing technical efficiency.

It is to be reiterated that the above tabulation represents the authors' best judgment on the structure of the programs that would be most advantageous for Puerto Rico, bearing in mind differential profitabilities and the fact that only a relatively small number of programs could be calculated in detail. The judgments that underlie the tabulation can, of course, be completely verified only by testing alternative programs.

The logic of the tabulation requires that acetylene be connected to the refinery before ammonia, and hydrogen cyanide (from methane) and ethylene before acetylene. The order of hydrogen cyanide and ethylene is arbitrary; in the production programs ethylene has been connected up first. Thus, the order of consideration of the basic petrochemicals is

THE FULL PROGRAMS 85

1. ethylene.
2. methane (for hydrogen cyanide).
3. acetylene.
4. ammonia.

There is one additional point that bears upon the scheduling of basic petrochemicals in the production programs detailed in Appendix B. This concerns the form of ammonia programming, and will be discussed with reference to Orlon J which has one of the more complicated ammonia schedules. The streams which serve as sources of ammonia are programmed in their order of preference, the total ammonia need in Orlon J being 101.50 MM lb/yr. Production of ammonia from the light platformer gas stream (see Table 10) is 43.91 MM lb/yr. The amount of H_2C_1 stream left over after the methane need for hydrogen cyanide is satisfied is 4% (0.040) of the total; this yields 70.01 × 0.04 = 2.80 MM lb/yr of ammonia. The amount of C_2—C_2C_3—C_3 stream left over after acetylene production is 14.4% of the total; this yields 98.04 × 0.144 = 14.12 MM lb/yr of ammonia. The balance of the ammonia requirement is produced from cycle oil. The ammonia table in the production program for Orlon (see Appendix B, Table B-7) shows the breakdown of the ammonia production from the various streams into the components of the streams. Thus the first row of the table shows the ammonia made from hydrogen, originating in the various streams; the second row shows the ammonia made from methane; the third row shows the ammonia made from ethylene; . . . etc. The numbers in the ammonia table are calculated by proportional reduction from the data of Table 10. Finally, the entries of each row are summed, thus yielding total ammonia made from hydrogen, methane, ethylene, . . . etc. These totals determine the scales of the ammonia activities.

In addition, fertilizer production must be considered. The fertilizer constraint restricts fertilizer production in Puerto Rico to the chemical equivalent of 80 MM lb/yr of ammonia. In most of the programs, the entire amount is programmed for production. As mentioned earlier, ammonium nitrate and urea are programmed for equal ammonia equivalent. Such a fertilizer program requires the production of 87.60 MM lb/yr of ammonium nitrate (40 MM lb/yr ammonia equivalent) and 69.00 MM lb/yr of urea (also 40 MM lb/yr of ammonia equivalent). Urea is produced from ammonia and carbon dioxide; ammonium nitrate is produced entirely from ammonia. For the latter purpose, part of the ammonia is oxidized to nitric acid (66.90 MM lb/yr of nitric acid production). This nitric acid is reacted with other ammonia to form ammonium nitrate. Accordingly, the fertilizer program is (8.760)(#55) + (6.900)(#56) + (6.690)(#43). This is referred to as the "standard" fertilizer program.[23] However, in a few cases (Orlon A, Orlon C, Orlon E, and Orlon H) fertilizer production has not been pushed to the limit. These cases permit comparison with otherwise equivalent programs with standard fertilizer production.

Now that we have discussed the various points relating to the construction of production programs, it may be helpful at this juncture to illustrate a program diagrammatically. For this purpose, the program Dacron A is chosen. This program typifies the refinery, petrochemical, fertilizer, and fiber activities of the full programs.

The diagrammatic sketch of Dacron A is given in Figure 2. Each square or rectangular block of the figure represents an activity of the Dacron A program. The number of each activity is indicated at the bottom of the respective block. The level at which each activity is to be operated is indicated by the number directly above the respective block. Thus; Figure 2 indicates (at the extreme right) that activity 4 (oil refinery prototype 4) is to be operated at a level of 1.000. From this activity flows a set of finished products such as gasoline and kerosene, which go to sales. However, there are also gas streams which emerge from the oil refinery. As portrayed in the upper part of Figure 2, ethylene is captured from these gas streams and used to produce ethylene glycol which in turn is combined with dimethyl terephthalate (based upon imported paraxylene) to yield Dacron polymer. From Dacron polymer, Dacron staple is obtained, the respective activity, activity 47, being operated at a level of 3.650.

The lower part of Figure 2 depicts how portions of the gas streams are directed to the production of ammonia via activities 31, 32, 33 and 34. (For example, part of the methane component of the gas streams is used in the operation of activity 32, ammonia from methane, at a level of 3.595.) The resulting outflows of ammonia are viewed as a pool from which the raw material for nitric acid, ammonium nitrate, and urea production are obtained.

---

[23] In some nylon programs, ammonium sulfate is obtained as a by-product of activity 69, the production of adipic acid from adiponitrile. In order to obtain the usual 80 MM lb/yr ammonia equivalent in these cases, the ammonium nitrate and urea production is reduced. The ratio of ammonium nitrate to urea is held constant.

THE FULL PROGRAMS 87

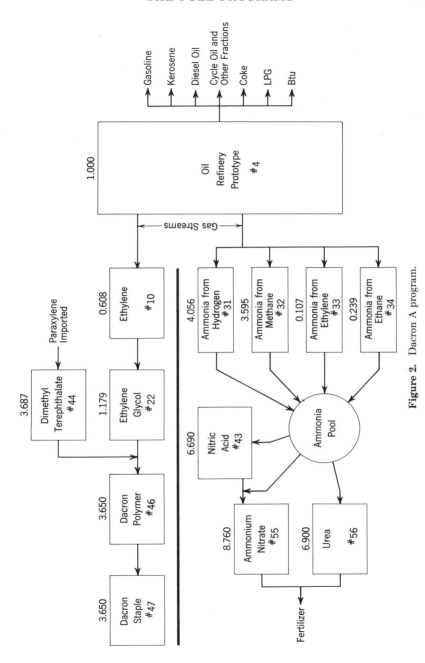

Figure 2. Dacron A program.

Before the discussion relating to production programs is concluded, two details remain to be cleared up. First, minimum-scale restrictions are violated in a number of programs. These violations are pointed out in notes appended to the individual programs in which they occur. The significance of the scale violations is discussed in Chapter 5 in connection with the evaluation of the programs.

Finally, a number of minor errors in programming and calculations were detected after the computations were made. These errors were considered too small to warrant recalculations or reconstruction of the tables.[24]

TOTAL INPUTS AND OUTPUTS, BY PROGRAM

The next step is the calculation of total inputs and outputs associated with each of the production programs. This calculation can be accomplished for any given program by multiplying all the inputs and outputs of each of the activities listed at the right of the production program, as detailed in Appendix B, by the corresponding multiple of unit level, which is listed beside each activity number. The results are summed by commodity (input or output) over all the listed activities to yield the total amounts for each input and output of the program.

---

[24] These errors are:

(a) The hydrogen by-product of chlorine production should have been given first preference in ammonia production wherever such hydrogen was available. This plan was not followed for Orlon F, Dynel A, Dynel B, Dynel C, and Dynel D.

(b) The by-product gases of ethane and propane cracking (for ethylene) should have been included among the streams shown in the above tabulation indicating order of preference of their use. These gases could be used for ammonia production (fourth preference) and possibly for acetylene production (third or fourth preference).

(c) An error was discovered in the ammonia program for Orlon F. While the total ammonia production is correct, the proportions of individual ammonia activities are off by about 1%. The above three errors have a very small effect on the end results, since the calculated advantages are insensitive to variation in the proportions of ammonia activities.

Also a dummy activity is missing from Table 3 which would transfer the pure by-product ethane of ethylene separation to the uses of mixed ethane. Since ammonia and acetylene activities are defined only for mixed (not for pure) ethane input, this omission creates errors in the commodity balances of mixed ethane and of pure ethane. These errors are equal in magnitude and opposite in sign. In the present study, the formal calculating procedure used for obtaining the final results cancels out these errors; consequently, no formal correction needs to be made.

THE FULL PROGRAMS

The production program for Orlon B, summarized in Table B-4, Appendix B, may be taken as an example. The first five entries recorded in the right-hand column constitute the relevant subprogram, as given in Table 9. The first entry indicates that activity 52, Orlon polymer production, must be operated at a scale of 3.650 times unit level. Therefore, each of the input and output amounts recorded in column 52 of Table 3 must be multiplied by 3.650. Thus, the operation of activity 52 requires 0.073 MMM lb of steam (0.020 times 3.650), 5.475 MM kw-hr of power (1.500 times 3.650), and 38.33 MM lb of acrylonitrile (3.650 times 10.500). The output of the activity is 36.5 MM lb of Orlon polymer (3.650 times 10.000).

The second entry in the right-hand column of Table B-4, Appendix B, indicates that activity 53, Orlon staple production, must be operated at a scale of 3.650 times unit level. Thus the input and output amounts in column 53 of Table 3 must be multiplied by 3.650. The steam required in the Orlon staple production stage is thus 1.825 MMM lb (0.500 times 3.650); etc.

The third entry in the Orlon B production program indicates that activity 41, acrylonitrile production must be operated at 3.833 times unit level. Individual input and output amounts associated with the operation of this activity can thus be obtained by multiplying the figures recorded in column 41 of Table 3 by 3.833. In a similar manner, the inputs and outputs of all the other individual activities required in the Orlon B program can be calculated.

Finally, the total amounts of any input associated with the entire Orlon B program can be calculated by summing the amounts of that input associated with the individual activities of the program. Thus the total steam required by the program is the sum of the steam requirements computed for the Orlon staple activity, the Orlon polymer activity, the acrylonitrile activity, etc. The amount is 4.885 MMM lb and is recorded in column 2 of Table 11. The final outputs of the program are computed in a like manner and are also recorded in column 2 of Table 11.

Each chemical intermediate produced within the complex in connection with the program appears as an output of one or more individual activities and as an input for one or more others. For such commodities, only *net* amounts appear as final inputs or outputs of the program. For example, 36.5 MM lb of Orlon polymer is an output of activity 52, but the same amount is required as an input for activity 53. Thus Orlon polymer does not appear at all as a final input or output in column 2 of Table 11.

90 INDUSTRIAL COMPLEX ANALYSIS

TABLE 11
TOTAL REQUIREMENTS AND YIELDS

	Orlon A 1	Orlon B 2	Orlon C 3	Orlon D 4	Orlon E 5	Orlon F 6	Orlon G 7	Orlon H 8
1 Crude Oil MMbbl	− 9.428	− 9.428	− 10.993	− 10.993	− 9.428	− 9.428	− 17.216	− 9.428
2 Gasoline, straight-run MMbbl	+ 1.300	+ 1.300	+ 1.516	+ 1.516	+ 1.300	+ 1.300	+ 2.374	+ 1.300
3 Gasoline, cracked MMbbl	+ 2.226	+ 2.226	+ 2.596	+ 2.596	+ 2.226	+ 2.226	+ 4.066	+ 2.226
4 Gasoline, reformed MMbbl	+ 1.486	+ 1.486	+ 1.733	+ 1.733	+ 1.486	+ 1.486	+ 2.713	+ 1.486
5 Gasoline, polymerized MMbbl	+ 0.444	+ 0.444	+ 0.518	+ 0.518	+ 0.444	+ 0.444	+ 0.811	+ 0.415
6 Naphtha, MMbbl								
7 Kerosene, MMbbl	+ 0.707	+ 0.707	+ 0.824	+ 0.824	+ 0.707	+ 0.707	+ 1.291	+ 0.707
8 Diesel Oil MMbbl	+ 0.896	+ 0.896	+ 1.045	+ 1.045	+ 0.896	+ 0.896	+ 1.636	+ 0.896
9 Gas Oil MMbbl								
10 Cycle Oil MMbbl	+ 1.980	+ 1.906	+ 2.309	+ 2.273	+ 1.980	+ 1.952	+ 3.615	+ 1.980
11 Heavy Residual MMbbl								
12 Coke and Carbon 10xMMlb	+ 4.033	+ 4.033	+ 4.703	+ 4.703	+ 4.033	+ 4.033	+ 7.384	+ 4.033
13 L.P.G. 10xMMlb	+ 14.006	+ 14.006	+ 18.512	+ 18.512	+ 15.739	+ 15.739	+ 28.409	+ 15.050
14 Hydrogen MMlb	+ 3.553	+ 3.553	+ 3.421	+ 3.421	+ 4.212	+ 4.212	+ 3.082	+ 1.640
15 Methane MMlb	+ 9.658	+ 9.658	+ 2.418	+ 2.418	+ 1.186	+ 1.186	+ 0.012	− 0.006
16 Ethylene (mixed) MMlb	+ 3.359	+ 3.359	+ 3.134	+ 3.134	+ 2.022	+ 2.022	+ 0.845	+ 0.950
17 Ethane (mixed) MMlb	+ 2.373	+ 2.373	+ 2.948	+ 2.948	− 11.062	− 11.062	− 54.030	+ 1.528
18 Propylene MMlb	+ 3.542	+ 3.542						+ 0.066
19 Propane MMlb								+ 0.001
20 Butylenes MMlb								
21 Butanes MMlb								
22 Pure Ethylene	− 0.012	− 0.012	− 0.009	− 0.009	− 0.011	− 0.011	− 0.012	
23 Pure Ethane MMlb	+ 0.004	+ 0.004	+ 0.003	+ 0.003	+ 13.033	+ 13.033	+ 55.127	
24 Steam MMlb	− 4.601	− 4.885	− 4.902	− 5.030	− 4.533	− 4.632	− 6.524	− 4.691
25 Power MMKWH	−115.032	−140.933	−128.097	−139.839	−260.851	−269.884	−378.981	−111.082
26 Fuel 10xMMMBTU	−304.131	−313.006	−297.792	−301.815	−295.372	−298.466	−323.618	−266.693
27 Salt MMlb					−111.401		−111.401	
28 Caustic Soda MMlb					+ 73.721		+ 73.721	
29 Chlorine MMlb					− 0.020	− 0.020	− 0.018	
30 Hydrochloric Acid 100% MMlb					+ 1.311	+ 1.311	+ 1.311	
31 Limestone MMlb					− 59.815	− 59.815	− 59.813	
32 Lime (hydrated) MMlb					+ 0.001	+ 0.001	+ 0.001	
33 Ethylene Oxide MMlb	+ 0.013	+ 0.013	+ 0.013	+ 0.013	+..0.013		+ 0.013	
34 Nitrogen MMlb	+265.948	+265.948	+265.948	+265.948				
35 Ethylene Glycol MMlb								
36 Ethylene Dichloride MMlb								
37 Acetylene MMlb								+ 0.002
38 Vinyl Chloride MMlb								
39 Ammonia MMlb	+ 0.001	− 0.003	− 0.008	− 0.008	− 0.003	− 0.002		− 0.012
40 HCN MMlb	+ 0.002	+ 0.002	+ 0.002	+ 0.002	+ 0.002	+ 0.002	+ 0.002	+ 0.002
41 Acrylonitrile MMlb	+ 0.005	+ 0.005	+ 0.005	+ 0.005	+ 0.005	+ 0.005	+ 0.005	+ 0.005
42 Methanol MMlb								
43 Sulphur MMlb								
44 Sulphuric Acid MMlb								
45 Nitric Acid MMlb	+ 0.031	+ 0.060	+ 0.054	+ 0.054	+ 0.055	+ 0.060	+ 0.173	+ 0.028
46 Paraxylene MMlb								
47 Dimethyl Terephthalate MMlb								
48 Dacron Polymer MMlb								
49 Dacron Staple MMlb								
50 Dacron Filament MMlb								
51 Dynel Polymer MMlb								
52 Acetone MMlb								
53 Dynel Staple MMlb								
54 Dynel Filament MMlb								
55 Orlon Polymer MMlb								
56 Dimethyl Formamide MMlb	− 8.395	− 8.395	− 8.395	− 8.395	− 8.395	− 8.395	− 8.395	− 8.395
57 Orlon Staple MMlb	+ 36.500	+ 36.500	+ 36.500	+ 36.500	+ 36.500	+ 36.500	+ 36.500	+ 36.500
58 Orlon Filament MMlb								
59 Ammonium Nitrate MMlb	+ 49.750	+ 87.680	+ 70.480	+ 87.680	+ 74.450	+ 87.680	+257.590	+ 43.580
60 Urea MMlb	+ 39.130	+ 68.960	+ 55.440	+ 68.960	+ 58.560	+ 68.960	+202.600	+ 34.270
61 Carbon Dioxide MMlb	− 29.348	− 51.720	− 41.580	− 51.720	− 43.920	− 51.720	−151.950	− 25.703

THE FULL PROGRAMS

TABLE 11
OF SELECTED INDUSTRIAL COMPLEXES

	Orlon J	Dynel A	Dynel B	Dynel C	Dynel D	Dynel E	Dynel F	Dacron A	Dacron B	Dacron C	Dacron D
	9	10	11	12	13	14	15	16	17	18	19
	- 9.428	- 9.428	- 9.428	- 9.428	- 9.428	- 10.786	- 9.428	- 9.428	- 9.428	- 9.428	- 9.428
	+ 1.300	+ 1.300	+ 1.300	+ 1.300	+ 1.300	+ 1.487	+ 1.300	+ 1.300	+ 1.300	+ 1.300	+ 1.300
	+ 2.226	+ 2.226	+ 2.226	+ 2.226	+ 2.226	+ 2.547	+ 2.226	+ 2.226	+ 2.226	+ 2.226	+ 2.226
	+ 1.486	+ 1.486	+ 1.486	+ 1.486	+ 1.486	+ 1.700	+ 1.486	+ 1.486	+ 1.486	+ 1.486	+ 1.486
	+ 0.415	+ 0.444	+ 0.444	+ 0.436	+ 0.436	+ 0.508	+ 0.415	+ 0.433	+ 0.428	+ 0.433	+ 0.428
	+ 0.707	+ 0.707	+ 0.707	+ 0.707	+ 0.707	+ 0.809	+ 0.707	+ 0.707	+ 0.707	+ 0.707	+ 0.707
	+ 0.896	+ 0.896	+ 0.896	+ 0.896	+ 0.896	+ 1.025	+ 0.896	+ 0.896	+ 0.896	+ 0.896	+ 0.896
	+ 1.887	+ 1.980	+ 1.980	+ 1.980	+ 1.980	+ 2.265	+ 1.980	+ 1.980	+ 1.980	+ 1.980	+ 1.980
	+ 4.033	+ 4.033	+ 4.033	+ 4.033	+ 4.033	+ 4.614	+ 4.033	+ 4.033	+ 4.033	+ 4.033	+ 4.033
	+ 15.050	+ 15.558	+ 15.748	+ 15.418	+ 15.421	+ 17.798	+ 15.050	+ 15.359	+ 15.270	+ 15.359	+ 15.270
	+ 1.640	+ 2.793	+ 2.540	+ 1.438	+ 1.193	+ 1.024	+ 0.610	+ 0.788	+ 0.845	+ 0.470	+ 0.524
	- 0.005	+ 3.536	+ 4.122	+ 0.038	+ 2.890	+ 10.181	+ 0.038	+ 15.068	+ 16.083	+ 9.549	+ 10.545
	+ 0.950	+ 1.106	+ 1.645	+ 4.932	+ 4.858	+ 0.819	+ 5.451	+ 6.948	+ 9.794	+ 6.715	+ 9.561
	+ 1.528	+ 1.059	+ 1.526	- 11.840	- 10.589	- 18.409	+ 9.805	+ 12.513	+ 17.812	+ 12.264	+ 17.564
	- 0.066			+ 2.085	+ 2.047		+ 2.331	+ 2.971	+ 4.290	+ 2.971	+ 4.290
	+ 0.001			+ 1.397	+ 1.372		+ 1.562	+ 1.991	+ 2.875	+ 1.991	+ 2.875
		+ 0.003	- 0.002	- 0.004	- 0.089	- 0.005		+ 0.003	+ 0.007	+ 0.003	+ 0.007
		+ 20.721	+ 12.130	+ 21.887	+ 22.039	+ 34.537		+ 18.356	+ 13.102	+ 18.356	+ 13.102
	- 5.021	- 4.235	- 4.304	- 4.354	- 4.275	- 4.540	- 4.401	- 4.113	- 4.192	- 4.547	- 4.342
	-142.383	-219.177	-178.899	-167.571	-200.343	-148.451	-148.253	-127.799	-128.750	-132.386	-163.048
	-277.014	-281.806	-284.055	-274.713	-273.522	-277.086	-272.244	-275.642	-277.528	-270.971	-272.855
		- 93.075	- 46.987	- 48.516	- 71.026	- 26.481	- 26.469		- 26.435		- 26.435
		+ 61.594	+ 32.106	+ 32.106	+ 47.003	+ 23.772	+ 17.516		+ 17.494		+ 17.494
									- 0.003		- 0.003
		+ 14.329	+ 14.329	+ 14.329					+ 0.311		+ 0.311
		- 29.927							- 14.185		- 14.185
		+ 0.003	+ 0.003		+ 0.003				- 0.005		- 0.005
			+106.352			+106.352		+ 80.172		+ 80.172	
								+ 0.001	+ 0.001	+ 0.001	+ 0.001
		- 0.003	- 0.003	- 0.003				+ 0.002			
	+ 0.002			+ 0.002				+ 0.005			
		+ 0.005	+ 0.005	+ 0.005	+ 0.005	+ 0.005					
	- 0.020	+ 0.001	+ 0.001	+ 0.006	+ 0.011	+ 0.011		+ 0.017	- 0.032	- 0.024	- 0.018	+ 0.004
	+ 0.002	+ 0.002	+ 0.002	+ 0.002	+ 0.002			+ 0.002			
	+ 0.005										
								- 2.522	- 2.522	- 2.522	- 2.522
										- 1.777	- 1.777
										- 0.002	- 0.002
	+ 0.060	+ 0.061	+ 0.061	- 0.020	- 0.020	- 0.020	- 0.020	+ 0.061	+ 0.061	+ 0.055	+ 0.055
								- 25.072	- 25.072	- 21.753	- 21.754
								+ 0.005	+ 0.005	+ 0.005	+ 0.005
								+ 36.500	+ 36.500	+ 36.500	+ 36.500
		- 10.950	- 10.950	- 10.950	- 10.950	- 10.950	- 10.950				
		+ 36.500	+ 36.500	+ 36.500	+ 36.500	+ 36.500	+ 36.500				
	- 8.395										
	+ 36.500										
	+ 87.680	+.87.600	+.87.600	+ 87.680	+.87.680	+ 87.680	+ 87.680	+ 87.600	+ 87.600	+ 87.600	+ 87.600
	+ 68.960	+ 69.000	+ 69.000	+ 68.960	+ 68.960	+ 68.960	+ 68.960	+ 68.960	+ 69.000	+ 69.000	+ 69.000
	- 51.720	- 51.750	- 51.750	- 51.720	- 51.720	- 51.720	- 51.720	- 51.720	- 51.750	- 51.750	- 51.750

It should be noted that the output scales recorded in the Orlon B production program for the basic petrochemicals and fertilizer activities reflect the allocation scheme for refinery gas raw materials discussed above, as well as the fact that the amounts of such raw materials are limited to the amounts associated with unit-level operation of refinery 4 and its associated gas-separation activity. For example, the ethylene required by the program amounts to 46.06 MM lb. The operation of the gas-separation activity (10) at unit level yields 16.10 MM lb of ethylene and 30.19 MM lb of ethane. This amount of ethane permits the operation of the ethane-cracking activity (14) at a scale of 2.340 times unit level, which yields 23.40 MM lb of ethylene. The remaining 6.56 MM lb of ethylene must be produced by operating the propane-cracking activity (15) at a scale of 0.656 times unit level. The total propane consumed in this manner amounts to 19.55 MM lb, and is transferred from the LPG pool by operating the dummy activity (13) at a scale of 1.955 times unit level.

After the methane requirements of the HCN activity and the requirements of the various gases for ethylene production have been met, the remainder of the refinery gas can be used for ammonia production. The remaining hydrogen permits the operation of activity 31 (ammonia from hydrogen) at a scale of 3.826 times unit level, which produces 38.26 MM lb of ammonia. In similar fashion, the remaining methane, ethylene, and ethane components of the refinery gas permit the production of a total of 23.76 MM lb of ammonia. However, an additional 34.61 MM lb of ammonia is required. It is obtained by operating the ammonia from cycle oil activity (38) at a scale of 3.461 times unit level.

Table 11 presents the total inputs and outputs associated with each of the nineteen selected programs involving Orlon, Dynel, or Dacron production. No total input or output figures appear for the nylon programs. The computation of the locational advantage amounts for the nylon programs proceeds by way of a short-cut method (to be discussed in Chapter 6) which bypasses the calculation of total inputs and outputs.

Examination of Table 11 indicates that, although all the input and output commodities associated with the activities of Table 3 are also recorded in Table 11, for any given program many of these commodities are represented by zero amounts. Either no inputs or outputs of such commodities are associated with any activity of the given program, or for any one commodity the amounts computed as outputs are just canceled by the amounts computed as inputs.

THE FULL PROGRAMS 93

In addition to the total inputs and outputs recorded in Table 11, total labor inputs of each program must be calculated.[25] Direct chemical–petroleum and textile labor inputs can be computed for each activity by using the labor factor and unit-level labor inputs given in Table 4.[26]

Total imported (chemical–petroleum-type) labor and total local (textile-type) labor can be derived from the direct labor amounts by the addition of labor inputs corresponding to such items as supervision, payroll overhead, indirect production cost, etc. The exact method and percentages used have already been discussed.[27] The total labor requirements of each major type are recorded in Table 12 for each of the specific full programs considered.

With the calculation of final inputs and outputs of the selected programs completed, the analysis proceeds to the evaluation for each program of the locational advantages and disadvantages of Puerto Rico compared to the U. S. mainland. The various steps of this evaluation constitute the subject matter of the next two chapters.

---

[25] Inputs of capital services need not be calculated because for such inputs no price differentials are assumed between Puerto Rico and the mainland.

[26] A computational procedure for obtaining direct chemical–petroleum labor inputs has been developed by Thomas Vietorisz (*Regional Programming Models and the Case Study of a Refinery–Petrochemical–Synthetic Fiber Industrial Complex for Puerto Rico,* unpublished Ph.D. dissertation, Massachusetts Institute of Technology, 1956, Chap. 13). The procedure is as follows:

1. The production programs, ordered in the numerical sequence of activities and with zeros inserted for missing activities, are laid out in vertical columns. Each program consists of a column of activity scales for production activities only. Scales of activities within the ethylene, acetylene, and ammonia group are recorded only as a single sum.

2. The labor factors of the production activities are laid out in a vertical column on the left edge of a separate sheet of paper, with the same spacing as the programs.

3. A "contracted" program is obtained by raising each activity scale to the corresponding labor factor. This can be done with the aid of a slide rule that has a log–log scale, although considerable care must be taken to avoid errors in reading. No fractional-power tables with sufficiently fine subdivisions are available. The most reliable method involves the use of logarithms.

4. The contracted program may be vectorially multiplied by the labor inputs (recorded for unit scale of the activities) in Table 4. A convenient way of doing this is to lay out all contracted programs in columns, side by side, and record the labor inputs for unit scale at the left edge of a separate sheet of paper, with identical spacing. Then, the two factors of each product can be read side by side, and the multiplication, performed on a desk computer, can be undertaken by proceeding vertically down the column.

[27] See Chapter 3, pp. 58–61

TABLE 12
TOTAL LABOR REQUIREMENTS OF SELECTED INDUSTRIAL COMPLEXES (100,000 MHR/YR)

		Textile labor	Chemical Petroleum labor
Orlon	A	53.04	22.96
	B	53.15	23.29
	C	52.98	22.75
	D	53.07	23.05
	E	53.93	25.80
	F	53.96	25.91
	G	54.17	26.58
	H	52.68	21.79
	J	52.80	22.17
Dynel	A	53.96	25.90
	B	53.46	24.31
	C	53.32	23.83
	D	54.08	26.31
	E	53.56	24.63
	F	53.34	23.90
Dacron	A	52.84	22.29
	B	53.59	24.73
	C	52.83	22.26
	D	53.58	24.69
Nylon	A	54.06	26.24
	B	54.17	26.58
	C	54.37	27.24
	D	53.74	25.19
	E	53.84	25.53
	F	54.04	26.19
	G	52.98	22.75
	H	53.13	23.21
	J	54.19	26.65
	K	53.80	25.41

Chapter 5

Derivation of Cost and Revenue Differentials by Commodity

As pointed out in earlier chapters, the first major step in the economic analysis of this study consists of a calculation of the locational advantages and disadvantages of various programs or complexes in Puerto Rico as compared to identical, efficiently located mainland complexes. The specific programs to be considered have been presented in detail in Chapter 4. Since the locational comparison is of identical programs, locational advantages and disadvantages can be computed by using for each commodity the relevant cost or revenue differential between the two locations. It is the object of this chapter to present an account of the estimates of the cost or revenue differential, or, more generally speaking, the price difference, for each commodity.

The cost (price) difference associated with the *purchase* of a commodity may well be different from the revenue (price) difference associated with the *sale* of the same commodity. For example, the import (purchase) price of fuel oil in Puerto Rico may be estimated as the Venezuelan price plus transport; the export (sale) price as the New York price minus transport. These two Puerto Rico prices are different. In contrast, in New York, the buying and selling price is substantially equal for any transaction that is not of extravagant size, since the New York market for fuel oil is so large that a transaction by a single refinery will leave the price unaffected. Thus, the price difference between Puerto Rico and New York for *fuel oil sale* is not identical with that for *fuel oil purchase*. Therefore in Table 13 price differences are recorded for both sale and purchase of each commodity. In most cases, the price differences can be regarded as invariant with the amounts sold or purchased within the present study. However, a few exceptions to this rule are encountered, due primarily to economies of scale in transport activities which manifest themselves in quantity discounts on transport rates. These will be discussed below, as the differentials which relate to each commodity are examined.

CRUDE OIL AND LIQUID REFINERY PRODUCTS

The price differences for crude oil and liquid refinery products depend primarily on transportation costs. A recent study of various reports and materials relating to tanker transportation is available in *The Location of Oil Refineries in the United States* by J. R. Lindsay.[1] These materials and Lindsay's analysis suggest that the published rates for the transportation of petroleum and its products are subject to violent fluctuations with momentary supply and demand conditions. Such rates apply only to a small fraction of the total shipments, since most shipments are undertaken either by tankers under long-term charter or by tankers owned and operated by integrated oil companies. In an attempt to arrive at a more realistic set of transport rates, Lindsay utilizes detailed first-cost, transport-capacity, and operating-cost estimates relating to tankers. He estimates that the cost of shipping a barrel of crude on the Galveston–New York run ranges between 1.73 and 2.07 cents per 100 barrel-miles (nautical).[2] In the present study, a figure of 1.8 cents per 100 barrel-miles (nautical) is used for the Galveston–New York run. For the somewhat shorter San Juan (Puerto Rico)–New

---

[1] Unpublished Ph.D dissertation, Harvard University, 1954.

[2] Lindsay's figures are consistent with calculations based on the following sources: (1) Report 440, Stevens Institute of Technology (around 1953), unpublished, by K. S. M. Davidson and H. W. McDonald; (2) U. S. Petroleum Administration for Defense, *Transportation of Oil,* Washington, 1951; (3) H. F. Robinson et al., "Modern Tankers," *Society of Naval Architects and Marine Engineers Transactions,* **56,** 422–471 (1948); (4) T. Koopmans, *Tanker Freight Rates and Tankship Building,* Netherlands Economic Institute, P. S. King & Son, Ltd., London, 1939; (5) First cost and operating cost data relating to tankers sailing under U. S. and Panamanian flags, for a U. S. shipping firm, obtained through the courtesy of Professor C. P Kindleberger of M.I.T.; (6) Standard Oil Co., *Register of Tank Vessels of the World,* rev. ed., 1949.

COMMODITY COST AND REVENUE DIFFERENTIALS 97

York run and for the considerably shorter Puerto La Cruz (Venezula)–San Juan run, the cost rates adopted are 1.87 and 2.40 cents per 100 barrel-miles (nautical), respectively.[3]

On the basis of these estimates, the following shipping rates for crude oil obtain:[4]

	Distance, nautical miles	Rate, ¢/100 bbl nautical miles	Transport cost, ¢/bbl
Galveston, Tex., to New York, N. Y.	1862	1.80	33.5
Galveston, Tex., to San Juan, P. R.	1715	1.80	30.9
Puerto La Cruz, Venez., to New York, N. Y.	1800	1.80	32.4
San Juan, P. R., to New York, N. Y.	1399	1.87	26.2
Puerto La Cruz, Venez., to San Juan, P. R.	550	2.40	13.2

Note: 1 nautical mile = 1.15 statute miles

Lindsay has also studied in detail the relative costs of shipping crude oil and finished products. His analysis indicates that there is no firm basis for anticipating differences in these costs. Therefore, in the present study, the same rates are applied to the shipment of finished products as to the shipment of crude.

---

[3] These estimates of cost rates for runs of varying lengths are based on the speed of the tanker and on the time spent in port at each end of the run, as well as on costs while in port and costs while at sea. Assumed speed is 13.9 knots (T–2 type tanker of about 16,500 deadweight tons capacity); stay in port is assumed to be 55 hours per round trip. The sources for these estimates are the ones cited in the previous footnote.

[4] These estimates assume average ships sailing under U. S. flag and average port facilities. It is, however, possible to reduce the disadvantage of the short 550-mile run to the vanishing point when: (1) special port facilities for rapid loading and unloading are available at both terminals; (2) ships are especially built for the short run with oversize pumping facilities that permit loading and unloading in as little as four hours; (3) approximately the same size ships can be used for the short run as for the long run (it has to be considered that the yearly carrying capacity of a ship of a given size rises sharply as the run becomes shorter); and (4) fast turn-around of ships is not prevented by administrative formalities or labor union objections. The factors generally working in favor of the short run are the following: (1) larger cargo capacity because of lower bunker requirements, and (2) no need to exceed most economical speed (about 12 to 13 knots) for the sake of fast deliveries (recently built long-haul ships often operate at 17 knots or even faster, thereby greatly increasing fuel consumption).

It should also be noted that the use of ships sailing under a Panamanian flag is legally possible on the Puerto La Cruz–San Juan run. Such use would reduce shipping cost by about 25% according to material contained in source 5 in footnote 2.

TABLE 13
PRICE (COST AND REVENUE) DIFFERENCES: PUERTO RICO LESS MAINLAND[1]

Part 1

Commodity No.	1	2	3	4	5	6	7	8	9	10	11	12	13	14	15	16
Sale,[2]	73.00	73.00	73.00	73.00	73.00	73.00	73.00	73.00	73.00	73.00	73.00	1.46	33.51	9.59	3.75	3.40
Purchase,[2]	143.00	309.00	309.00	309.00	309.00	309.00	309.00	309.00	309.00	143.00	143.00	4.60	5.53	750.00	90.00	45.00

Part 2

Commodity No.	17	18	19	20	21	22	23	24	25	26	27	28	29	30	31	32
Sale,[2]	3.51	3.30	3.40	3.28	3.35	3.40	3.51	235.35	0.71	1.57	0	-2.93	-5.53	-13.22	0	-4.60
Purchase,[2]	45.00	5.53	5.53	5.53	5.53	45.00	45.00	235.35	6.8	1.57	0	5.82	5.53	13.22	0	4.60

Part 3

Commodity No.	33	34	35	36	37	38	39	40	41	42	43	44	45	46	47	48
Sale,[2]	-5.53	0	-3.82	-4.25	-45.00	-5.53	-5.53	-53.3	-6.8	0	-4.60	-4.25	-6.31	-3.82	-6.80	-6.80
Purchase,[2]	5.53	45.00	3.82	4.25	45.00	5.53	5.53	5.53	6.8	0	4.60	4.25	6.31	3.82	6.80	6.80

TABLE 13 (Con't)

Part 4

Commodity No.	49	50	51	52	53	54	55	56	57	58	59	60	61
Sale,[2]	-18.9 X	-18.9	-6.80 X	-3.82	-18.9	-18.9 X	-6.80	-3.82	-18.9	-18.9	+7.59	+10.12	0
Purchase,[2]	18.9	18.9	6.80	3.82	18.9	18.9	6.80	3.82	18.9	18.9	7.59	10.12	0

Part 5

Commodity No.	62	63	64	65	66	67	68	69	70	71	72	73	74	75	76
Sale,[2]	-3.82 X	-3.82	-3.82 X	-6.80 X	4.60 X	3.82	-3.82	-6.80 X	-3.82	-12.80	4.60	-6.80 X	-6.80 X	-18.9	-18.9
Purchase,[2]	3.82	3.82	3.82	6.80	-4.60	-3.82	3.82	6.80	3.82	12.80	4.60	6.80	6.80	18.9	18.9

[1] The differences are expressed in units of $1000. Those which are actually used in the programs of the present study are underlined. It is noted that individual commodities are either purchased by all programs, or sold by all programs. This characteristic is due to the structure of the programs, not to the price differences. Some price differences not utilized in the programs of the study have been omitted.

The commodities which occur with zero sales or purchases in all programs of the study are marked with an X at the dividing line of sales and purchases. The price differences of such commodities are not necessary for obtaining the results of the present study.

[2] See discussion in text for the relevant commodity units which differ from commodity to commodity.

Regarding the price structure for crude oil, Lindsay finds that the prices of Gulf Coast crude and of Venezuelan crude are equal in the New York area[5] and that refineries in the area are actually charging both kinds of crudes. On the basis of this finding, and taking all production and transport costs into account, Lindsay calculates that, for serving the Eastern Seaboard market, it is more profitable to refine Gulf Coast crude at a Gulf Coast location and to transport the products to New York than to refine Venezuelan crude in New York.[6] However, because there is a greater profit in producing Venezuelan crude than in producing Gulf crude, it makes sense for an integrated crude-producing and refining company to import Venezuelan crude and to refine it in New York. (The possibility of locating a refinery in Venezuela and importing the refined products is precluded by the present tariff structure.) If the price of crude oil were determined competitively, the price of Venezuelan crude in New York would have to be somewhat lower than the price of Gulf crude transported to New York, in order to compensate for the difference in the profitability of refining operations in these two locations. A rough estimate based on Lindsay's data indicates that the price spread so determined would be about 2¢/bbl of crude.[7]

In the present study, it is assumed that the prices of Gulf Coast and Venezuelan crude are the same in New York, as is actually so. *All results are obtained on this basis.* However, the possibility of

---

[5] Crudes of different qualities (re: product yields and impurities, especially sulfur) are subject to appropriate discounts or premiums which are just sufficient to eliminate differences in the profitability of refining these crudes.

[6] This conclusion holds for all but two of the approximately two dozen refinery prototypes Lindsay has investigated in detail. The two odd prototypes are unusual in the arrangement of their operations and not representative of refineries in general. The differences in profitability are, however, never more than 10 to 15% of total profits, and in many cases considerably less; thus no major change in future locational patterns can be expected.

[7] This implies that under perfect competition no Gulf crude would be refined in New York, and that the price of Venezuelan crude in New York would be 2 cents below the price of crude at the Gulf plus transport cost.

The price spread for crude oil required to offset the different profitabilities of refining at the Gulf or in New York varies considerably among the two dozen different refinery prototypes examined by Lindsay. The price spread ranges from about 7¢/bbl to −3¢/bbl. The 2¢/bbl spread is an approximate average of four basic prototypes (Lindsay's III*a*, alternate III*a*, III*b*, and alternate III*b*) which embody topping, catalytic cracking, and catalytic polymerization, but not re-forming, coking, visbreaking, or lube-oil manufacture.

a 2¢/bbl discount on Venezuelan crude will be investigated briefly in the critical evaluation of the relevance of these results.[8]

A potential refinery in Puerto Rico is to be compared with the most favorable alternative location on the mainland. Assuming that the prices of Venezuelan and Gulf crudes are equalized in New York, the most favorable location on the mainland is the Gulf Coast. The profit advantage for a Gulf Coast location as compared with a New York location is even greater when not refineries alone but also integrated refinery–petrochemical–polymer operations are considered. This is so for two reasons. One, the best mainland location for synthetic fiber production from its immediate chemical raw material, the synthetic polymer, is the Southeast (see below); consequently, in the case of a New York location, the petrochemical and polymer content of Venezuelan or Gulf crude has to be shipped first to New York and then back to the Southeast, whereas, in the case of a Gulf Coast location, there is only one short shipment of polymer to the Southeast. Two, petrochemical–polymer operations add considerably to fuel consumption; some fuel prices are significantly lower in the Gulf Coast, and this increases the advantage of a Gulf Coast location.

Accordingly, in the calculation of price differences for refinery, petrochemical, and polymer commodities, Puerto Rico is compared with the U. S. Gulf Coast. To the extent that refinery and petrochemical operations in Puerto Rico are a substitute not for Gulf Coast but for Eastern Seaboard locations, the advantage of Puerto Rico is understated. This is in line with the policy of this study to secure firm minimum estimates rather than uncertain estimates of average gains.

In line with the foregoing considerations, the price difference for the *purchase of crude oil* is determined as follows:

(1) P_{GC} = Price at U. S. Gulf Coast = New York price minus transport cost of 33.5¢/bbl = $(P_{NY} - 33.5)$;

---

[8] Since it is unrealistic to assume perfect competition, the 2¢/bbl discount (when employed) is applicable only to the internal accounting price of crude oil which an integrated oil company may assign to the quantities of crude transferred from its producing department to its refining department. The purpose of such a discount would be to ensure a better comparison between alternative refinery locations at the Gulf, in New York, or in Puerto Rico. The discount would, however, typically not be available to an independent refiner who is buying his crude at a price based on the equalized price in the New York market. Such a refiner would, for example, have to pay the equalized New York price minus 32.4¢/bbl for crude f.o.b. Puerto La Cruz, Venezuela.

(2) P_{PR} = Price in Puerto Rico = Puerto La Cruz price plus transport cost of 13.2¢/bbl = $(P_{PLC} + 13.2)$. However, the price in Puerto La Cruz is the New York price minus transport cost of 32.4¢/bbl = $(P_{NY} - 32.4)$.[9] Therefore, $P_{PR} = (P_{NY} - 32.4 + 13.2) = (P_{NY} - 19.2)$.

(3) Finally, $(P_{PR} - P_{GC}) = (P_{NY} - 19.2) - (P_{NY} - 33.5) = 14.3$¢/bbl.

The price difference for the *sale of crude oil or other liquid refinery streams (gasoline, naphtha, kerosene, Diesel oil, gas oil, cycle oil, and heavy residual)* is calculated likewise:

(1) $P_{GC} = (P_{NY} - 33.5)$.

(2) Price in Puerto Rico equals price at the market in New York minus transport cost from Puerto Rico to New York of 26.2 cents; i.e., $P_{PR} = (P_{NY} - 26.2)$.

(3) $P_{PR} - P_{GC} = (P_{NY} - 26.2) - (P_{NY} - 33.5) = 7.3$¢/bbl.

The sale of crude oil is not included in any program for Puerto Rico (see previous chapter); however, the formal price difference for the sale of this commodity is the same as for the sale of other refinery products.

For the purchase of fuel oil, i.e., heavy residual or cycle oil, the relevant Puerto Rico price is the import price from Venezuela; for the purchase of other liquid refinery products, the relevant Puerto Rico price is the import price from the Gulf, since the tariffs on these products are prohibitive. Hence, for the *purchase of fuel oil*, the price difference is the same as for crude oil, i.e., 14.3¢/bbl. For the *purchase of other liquid refinery products*, the Puerto Rico price is $(P_{GC} + 30.9)$; i.e., the price difference $(P_{PR} - P_{GC}) = (P_{GC} + 30.9) - (P_{GC}) = 30.9$¢/bbl.

In order to obtain the values shown in Table 13, all cents per barrel figures are converted to units of thousands of dollars per million barrels.

---

[9] This price is the price after the payment of import duties. The duties are absorbed by the crude oil producer in order to make his price competitive in the New York market. The amount of Venezuelan oil imported to the United States by the integrated crude-oil-producing and refining companies is determined not by considerations of profit but by considerations of public relations and legislative pressure. Although there is no formal quota in effect for imports at the time of this writing, the threat of the enactment of such a quota holds back the integrated oil companies from importing considerably more than they do. Consequently, the quantity imported can be regarded as fixed, and the price of Venezuelan crude oil is determined like a rent. (However, this analysis would cease to be valid if world demand were to bid up the Venezuelan price above the level determined by the New York price.)

The same statements are applicable with respect to duties on Venezuelan fuel oil imports.

GASEOUS FUEL PRODUCTS

It has already been indicated that the price difference on fuel oil sales is 7.3¢/bbl, a difference equivalent to 1.17¢/MM Btu.[10] The price difference per million Btu on the *sale of gaseous fuels* is, however, different.

At the Gulf Coast, liquid fuels are considerably more valuable than gaseous fuels. Gaseous fuels not only are produced by petroleum refineries, but also are available as natural gas. The region has a surplus of both liquid and gaseous fuels which are sold primarily on Eastern Seaboard markets. Whereas the transport cost of fuel oil via tanker from Galveston to New York is only $33.5/6.25 = 5.4$¢/MM Btu, the transport cost of fuel gases (primarily natural gas) via pipeline is estimated at 19.9¢/MM Btu.[11] Thus, at the Gulf Coast, gaseous fuels are estimated to sell at a discount of $19.9 - 5.4 = 14.5$¢/MM Btu relative to liquid fuels.

No such price difference between gaseous and liquid fuels is anticipated for Puerto Rico. It is assumed that, if any surplus of gaseous fuels is produced by a refinery–petrochemical–fiber complex, such surplus will not exceed local demand in Puerto Rico. Thus, there is no need to transport the gases to a distant market. Furthermore, it is assumed that surplus refinery gas and fuel oil can be used interchangeably in satisfying local fuel demand (on a Btu-equivalent basis); consequently, the price of fuel gases is determined by the price of fuel oil.

The price of fuel oil in Puerto Rico depends on whether there is an exportable surplus or a deficiency requiring imports. (The import of gases for use as a fuel is precluded by their very high overseas transport cost.) In the present study, this problem is handled on a simplified basis. All programs surveyed contain a refinery, and likewise all programs provide an exportable fuel oil surplus. Consequently, the relevant fuel price for Puerto Rico is the fuel oil export price. It is assumed that any local fuel sales will be undertaken at this same price: i.e., that no advantage is taken of any monopoly position in the local fuel market. The export price for

---

[10] Following Lindsay, *op. cit.*, a conversion factor of 6.25 MM Btu/bbl of fuel oil is assumed.

[11] This estimate is based on large-diameter (26- to 30-in.) pipelines used at a high load factor. Under such conditions, the transport cost is 1.4¢/1000 cu ft per 100 miles. Assuming 1000 Btu/cu ft heating value and 1420 miles distance, the transport cost is 1.4×14.20¢/MM Btu = 19.9¢/MM Btu (Walter Isard and Eugene W. Schooler, *Location Factors in the Petrochemical Industry*, U.S. Department of Commerce, Office of Technical Services, 1955, p. 18.

fuel oil is the New York price minus transport cost of 26.2¢/bbl (containing 6.25 MM Btu), or the New York price minus 4.2¢/MM Btu. Consequently, the relevant price difference on the *sale of gaseous fuels* between Puerto Rico and the Gulf Coast is $(P_{PR} - P_{GC}) = (P_{NY} - 4.2) - (P_{NY} - 19.9) = 15.7$¢/MM Btu. This compares with 1.17¢/MM Btu which is derived at the beginning of this section for the sale of liquid fuels. The price difference of 15.7¢/MM Btu is used for the calculation of price differences for the net outputs of hydrogen, methane, ethylene, ethane, propylene, propane, butylene, butane, pure ethylene, and pure ethane. It is assumed that the outputs of these gases which are not utilized by the complex for chemical purposes can be used only for fuel. The entries in Table 13 for the sale of these commodities are based on their fuel values.[12] The price differences for their purchase are based on an entirely different set of considerations and are discussed below.

FUEL INPUTS: STEAM AND POWER

Fuel inputs for the complex are handled on the same simplified basis, i.e., on the assumption of an exportable fuel oil surplus. Since gas, the cheapest fuel at the Gulf, will invariably be used for fuel needs at the Gulf, while all fuels have the same price in Puerto Rico, the relevant price difference for fuel inputs is the same as the price difference for gaseous fuel sales, i.e., 15.7¢/MM Btu. (Certain calculations based on this difference will be subject to an adjustment because of the split-location patterns on the mainland, as will be discussed below.) The price differences for the *inputs of steam and power* are also based on the same 15.7¢/MM Btu price differential, on the assumption that the generation of steam requires 1500 Btu/lb and the generation of power requires 4500 Btu/kw-hr.

[12] These fuel values (gross heating values per pound of gas) are in Btu per pound:

Hydrogen	61,000	Propylene	21,060
Methane	23,920	Propane	21,690
Ethylene	21,650	Butylenes	20,800
Ethane	22,350	Butanes	21,300

Source: For hydrogen: Shreve, *op. cit.*, p. 100.
For all other gases: W. L. Nelson, *Petroleum Refinery Engineering*, 3rd ed., McGraw-Hill Publishing Co., 1949, pp. 150-153.
The figures given for butylenes and butanes represent the average values for (a) *n*-butene-(1), *n*-butene-(2) and isobutylene; (b) for *n*-butane and isobutane.

COMMODITY COST AND REVENUE DIFFERENTIALS 105

The Btu requirement for *steam* generation depends on the pressure of steam.[13] The heat content of saturated steam is at a maximum of 1204 Btu/lb at 400 psig pressure, and is slightly lower both at pressures below and at pressures above this level. Assuming a feed-water temperature of 70° F (with a heat content of 38 Btu/lb) the maximum heat of evaporation for obtaining saturated steam at 400 psig is 1166 Btu/lb. The figure of 1500 Btu/lb, accordingly, allows for a boiler efficiency of about 80% under these conditions. If steam is generated at a higher pressure, the heat of evaporation is slightly less; on the other hand, the total heat content of the steam may be increased by applying superheat. Thus, the figure of 1500 Btu/lb is a reasonable representation of the various cases that may arise. On this basis, the price difference per million pounds of steam is 1500 × 15.7¢ = 23,535¢, or $235.35 M/MMM lb.[14]

On the generation of *power,* the figure of 4500 Btu/lb represents a situation in which power is obtained as a by-product of process steam. In refinery and chemical processes, large amounts of low-pressure steam are consumed. By-product power can be obtained by generating steam at a high pressure, passing it through a steam turbine, and using the low-pressure exhaust steam for process. In such a joint production of process steam and power, the *extra* heat input required, over and above the Btu need for process steam, is about 4500 Btu/kw-hr.[15] On this basis, the price difference per million kilowatt-hours is 4500 × 15.7¢ = $0.71 M.

The amount of power that can be obtained in this way may range anywhere between 20,000 and 100,000 kw-hr/MM lb of process steam.[16] The exact amount of power obtainable depends on the type of power plant. Steam-turbine power plants generally furnish lower amounts of by-product power than gas-turbine power plants equipped with waste heat boilers. The best yields obtainable by the former are in the 60,000 to 70,000 (kw-hr/MM lb steam) range; by the latter in the 90,000 to 100,000 range. The yields of steam-turbine by-product power plants are strongly related to two variables: initial steam pressure, and pressure of

---

[13] The discussion here follows Isard and Schooler, *op. cit.,* p. 15, footnote 2.

[14] In this multiplication, as with others, minor errors develop from the rounding of numbers.

[15] From John H. Perry, ed., *Chemical Engineers' Handbook,* 3rd ed., McGraw-Hill Publishing Co., 1950, p. 1630, Table 4.

[16] B. Wilson, "Should Your Plant Produce Power," *Chemical Engineering,* **60,** 176 (Mar. 1953), Fig. 1.

exhaust steam supplied to process. The highest yields are obtained when initial steam pressure is high and exhaust steam pressure is low.

As in other processes, significant economies of scale are encountered in power generation. Even though the fuel economy of a by-product power plant is three times greater than that of a main-product power plant, the total unit cost per kilowatt-hour may be higher for a by-product power plant when the scale is small. At the scales of 9 MM kw-hr/yr (1000 kw) and 90 MM kw-hr/yr (10,000 kw), the cost of by-product power from a steam-turbine power plant at a 75% load factor is 1.4 and 0.6¢/kw-hr, respectively. The corresponding figures for main-product power plants are 3.0 and 1.3¢/kw-hr.[17] The cost for gas-turbine power plants is estimated to be slightly lower, but there is considerably less industrial experience with gas than with steam turbines.[18]

For the purposes of the present study, it is assumed that

(a) The amount of by-product power available is always sufficient to supply the power needs of the complex. In fact, the highest ratio of power to steam encountered in the programs calculated in detail is that for Orlon F, namely, 58,300 kw-hr/MM lb of steam. (See Table 11.)

(b) The scale of the power plant is large enough to eliminate significant diseconomies of scale. In the programs calculated in detail, the smallest power requirement is 111 MM kw-hr/yr, or about 12,300 kw.

(c) Any excess by-product power capacity is left unutilized. This understates the potential advantage of Puerto Rico. (At present, part of the power generated for public use is based on fuel oil.[19] Whatever portion of the latter can be replaced with by-product power represents a fuel saving of about 9000 Btu/kw-hr; this could be added to the advantage of a complex in Puerto Rico

---

[17] Wilson, op. cit.

[18] For a discussion of gas turbines in by-product power generation, see Wilson, op. cit. Gas turbines can also provide compressed air or gases very economically; this permits considerable savings when gas turbines are used in such processes as air separation, low-temperature gas fractionation, ammonia manufacture, nitric acid manufacture, and other high-pressure processes. Since these applications of the gas turbine are new and unproved, they have not been included in the present study. For details, see Benjamin Miller, "Gas Turbines for Process Use, I–II," *Chemical Engineering*, **62**, 175–180 (Jan. 1955) and 187–192 (Feb. 1955).

[19] *Facts for Businessmen*, Puerto Rico Economic Development Administration, San Juan (not dated, approximately 1952), pp. 29, 61–62.

COMMODITY COST AND REVENUE DIFFERENTIALS 107

at the usual price difference on fuel cost, namely 15.7¢/MM Btu.) Whenever power needs exceed the amount of by-product power available, the fuel input per kilowatt-hour is approximately trebled,[20] and the price difference applicable to the main-product power becomes $2730/MM kw-hr.

LPG AND COKE

The price difference for the *sale of liquefied petroleum gas* (LPG) is taken to be the same on a Btu basis as the price difference on the sale of other gases.[21] This price difference results from the assumption that the premium obtained for LPG over alternative fuels is identical at the Gulf Coast and in Puerto Rico.[22] (It is expected that the entire LPG output of the complex can be sold locally in Puerto Rico, since domestic gas distribution facilities do not exist in most of the island, and since the market for fuel will expand steadily as living standards rise.)

The *purchase of LPG* does not occur in any of the programs that have been calculated. A price difference for this activity is obtained in a later section, based on overseas transport cost.

The price difference for the *sale of coke* is difficult to estimate without extensive investigation of the pattern of markets and of transport costs. In general, however, the more important uses for petroleum coke are: (1) as an industrial fuel, (2) as a raw material for carbon electrodes, especially for aluminum manufacture, and (3) as a raw material for electrothermal chemical processes, such as the manufacture of graphite, silicon carbide, and calcium carbide.[23]

---

[20] Perry, *op. cit.*, p. 1630, Table 4. Some of the figures quoted are: Diesel plant, 11,500 Btu/kw-hr; natural gas engine plant, 14,000 Btu/kw-hr; central-station steam plants, average, 16,000 Btu/kw-hr; small condensing industrial steam plant, 20,000 Btu/kw-hr.

[21] The heating value of LPG is assumed to be 4.250 MM Btu/bbl. This figure is based on an estimated density of 199 lb/bbl and heating value of 21,300 Btu/lb. The latter figures are averaged from the major components of LPG: propane and butanes. See Nelson, *op. cit.*, pp. 150–153.

[22] LPG is assumed to sell at a premium price over competing fuels, because it can be burned in places where gas-distribution facilities are lacking, and because it is superior to fuel oil as a domestic fuel.

[23] For a discussion of the markets for petroleum coke, with numerical estimates by categories for the years 1952, 1953, and 1954, see George Weber, "Petroleum Coke," *Oil and Gas Journal,* **52,** 151–154 (Mar. 22, 1954).

Thus, on the one hand, coke competes with other fuels—primarily fuel oil and natural gas; on the other hand, it serves a specialty market which is locationally attracted to cheap power. For fuel use the influence of the New York market is dominant; for other uses the markets are in the Pacific Northwest, in the TVA area, and in several new locations of aluminum manufacture (east Texas, the Ohio Valley, east and west Canada). Since it is difficult to predict the relative importance of these various markets, it was arbitrarily decided to assign to the sale of coke a price difference equal on a Btu basis to the price difference on the sale of fuel oil. The result is a difference of 1.46¢/100 lb of coke, or $0.146 M/MM lb.[24]

The price difference for the *purchase of coke* is simply based on import of coke to Puerto Rico from the Gulf. This is discussed at a later point.

In the foregoing, a simplified method for handling fuel outputs and inputs has been presented. This method depends on the special assumption of an exportable fuel oil surplus in Puerto Rico. The price difference of 15.7¢/MM Btu was found to apply to the outputs of gaseous fuels and LPG and to the inputs of all fuels for process heat, steam, and power. The price difference of 1.17¢/MM Btu was found to apply to the outputs of liquid fuels and coke. The simplified approach is applicable to all those programs that were calculated in detail in the present study.

However, the possibility of price differences based on the import of fuel oil to Puerto Rico must also be recognized. This situation would characterize programs involving a fuel deficit (such as are discussed in the following chapter). In such programs, Puerto Rico's disadvantage relative to the Gulf on *fuel oil purchase,* as derived above, is 2.29¢/MM Btu (14.3¢/bbl). Since Gulf Coast activities would always burn low-cost gas, Puerto Rico's disadvantage on fuel inputs would be 2.29 cents plus an additional 14.5 cents (fuel oil–gas differential at the Gulf) or 16.79¢/MM Btu. The same price difference would apply to any *output* of gaseous fuels in the program. The corresponding price difference on inputs

---

[24] The heating value of coke, taken from Perry, *op. cit.,* p. 1629, Table 2, is 12,500 Btu/lb, or 1.25 MM Btu/100 lb. The applicable price difference based on fuel oil is 1.17¢/MM Btu; thus, per hundred pounds the price difference for coke is 1.25 × 1.17 = 1.46¢.

COMMODITY COST AND REVENUE DIFFERENTIALS 109

of steam, at 1500 Btu/lb, is $251.85/MM lb, and on inputs of power, at 4500 Btu/kw-hr, $750.00/MM kw-hr.[25]

LIQUID, SOLID, AND GASEOUS CHEMICALS

The price differences for the purchase and sale of many commodities are based simply on the transport costs incurred in importing or exporting them. With respect to the chemical commodities which appear in the selected programs of this study, it is reasonable to assume that any necessary imports to Puerto Rico would come from the New York area or from the Gulf Coast. Most of the chemicals are produced at present in at least one of these locations (e.g., chlorine, ammonia and various other petrochemicals, and sulfur).[26] In other cases, the nature of the raw materials, markets, and production processes makes it likely that either the Gulf Coast or the New York area will be a favorable location in the long run.[27]

For chemicals produced in Puerto Rico for export to the United States, the markets are also likely to be primarily in the New York

---

[25] For further discussion of the fuel valuation problem and for a more generalized method of determining relevant price differences for fuel inputs and outputs, see Thomas Vietorisz, *Regional Programming Models and the Case Study of a Refinery-Petrochemical-Synthetic Fiber Industrial Complex for Puerto Rico*, unpublished Ph.D. dissertation, M. I. T., 1956, Chap. 12.

[26] See the location maps in Faith, Keyes, and Clark, *Industrial Chemicals*, 2nd ed., John Wiley & Sons, 1957.

[27] For example, present dimethyl terephthalate production in the United States is tied to the location of the production of Dacron fiber. But the existing production facilities were built when main consideration was given to products, process, and market development, and not to production cost. Consequently, the present plants cannot be taken to represent a long-run equilibrium locational pattern.

Dimethyl terephthalate is based on paraxylene. The chief future source of this chemical is refinery naphtha. Even though mixed xylenes are also obtained from coke-oven operations as a by-product, the amount so produced is very closely tied to coke, and hence to steel production. Consequently, as demand for xylenes increases, it is likely that most of the expansion will take place in petrochemical xylene production. The best location for such production in the United States is the Gulf Coast (see above, this chapter). As a result, both the location of paraxylene supply and the availability of cheap fuel draws dimethyl terephthalate production to the Gulf. Nevertheless, the New York area is a strong competitor, because refineries located in this area have an advantage in importing crudes from Venezuela and the Middle East; a good deal of petrochemical capacity is also located in this area.

area or at the Gulf Coast. (Some exceptions are discussed below.)
For all these commodities the price differences adopted in this study reflect overseas transport cost. For imports, the price in Puerto Rico is taken to exceed the mainland price by the amount of the transport cost. For exports, the price in Puerto Rico is assumed to be smaller than the mainland price by an amount equal to the transport cost.

For certain of the commodities, the overseas transport cost is derived from a calculation of the distance shipped multiplied by an appropriate transport rate.[28] For others, however, the estimated transport cost is taken directly from quotations by shipping associations.

Liquid chemicals constitute the principal group of commodities for which transport costs are calculated by the multiplication of rate and distance. The rates used in this study are adapted from detailed estimates by Isard and Schooler, based on the cost of building and operating chemical tanker fleets. These authors distinguish three classes of chemical products, from the standpoint of shipping costs: ordinary chemicals, corrosive chemicals, and chemicals shipped under pressure. The corresponding overseas transport rates, in cents per ton-mile, are 0.45, 0.50, and 0.65.[29] These rates are used to compute the transport costs which are recorded in the first column of Table 14 and which apply to chemicals shipped in quantities of more than 10 million pounds per year. The other figures of Table 14 relate to smaller shipments and are calculated by adding an estimated diseconomy of scale adjustment to the above rates. For shipments of 5 to 10 MM lb/yr, the rates are increased by 50%; for shipments of less than 5 MM lb/yr, the rates are increased by 100%.[30]

---

[28] The approximate average distance from Puerto Rico to the two mainland markets is taken to be 1700 statute miles. The San Juan–New York and the San Juan–Galveston distances are 1399 and 1715 nautical miles, respectively. Their average, 1557 nautical miles, equals 1780 statute miles. The use of the somewhat lower figure of 1700 statute miles in computing price differences assigns more weight to the New York market than to the Gulf Coast market.

[29] Isard and Schooler, *op. cit.,* p. 28. In this reference, rates are estimated for ships of 5000 tons and for those of 10,000 to 12,000 tons. For the present study, the figures applicable to the smaller-ship size are adopted. All rates apply to long intercoastal shipments and do not include terminal or handling charges.

[30] A check on the figures of Table 14 is obtained by a comparison with selected rates from the "United States Atlantic and Gulf–Puerto Rico Tariff," *Outward Freight Tariff no. 6,* issued by J. W. de Bruycker, agent, 8–10 Bridge Street, New York, N. Y., November 2, 1951, with later supplements. The lowest rate in this tariff for a non-

COMMODITY COST AND REVENUE DIFFERENTIALS 111

TABLE 14
LIQUID CHEMICALS: ESTIMATED OVERSEAS TRANSPORT COST
BETWEEN PUERTO RICO AND U. S. MAINLAND

	Volume Shipped		
	Above 10 MM lb/yr	5-10 MM lb/yr	Below 5 MM lb/yr
Class of Chemical	Transport Cost $M/MM lb		
1. Ordinary Chemicals	3.82	5.73	7.64
2. Corrosive Chemicals	4.25	6.38	8.50
3. Pressure Chemicals	5.53	8.30	11.60

Of the liquid chemicals appearing in the programs of the present study, ethylene glycol, methanol, paraxylene, acetone, dimethyl formamide, benzene, cyclohexane, cyclohexanone, furfural,[31] tetrahydrofuran, adiponitrile, and hexamethylene diamine are classed as ordinary chemicals; ethylene dichloride, sulfuric acid, nitric acid, hydrochloric acid, and caustic soda are classed as corrosive chemicals; propylene, propane, butylene, butane, LPG, chlorine, ethylene oxide, vinyl chloride, ammonia, hydrogen cyanide, and butadiene are classed as pressure chemicals. The third group consists of gases which, under moderate pressure, condense into liquids.[32] Nitric acid, hydrochloric acid, and caustic soda are recorded throughout the study on a water-free basis; however, for the pur-

---

corrosive liquid is quoted on fuel oil in bulk (shipment over 1500 bbl = 525,000 lb), $1.75 per 42-gal barrel (December 11, 1951); i.e., $5000/MM lb (@ 350 lb/bbl). Kerosene is next with a rate of about $5500/MM lb (December 11, 1951). Alcohol and gasoline take a rate of $9700 and $12,000/MM lb, respectively (November 3, 1952); ammonia (anhydrous), $9000/MM lb (April 27, 1953); propane and butane (in tanks, on skids, strapped) $11,900/MM lb (November 3, 1952). The orders of magnitude agree with Table 14; the tariff rates are generally higher, reflecting the commodity rate policy of the common carrier. The lowest figure (for fuel oil) agrees well with the rate given in Table 14 for ordinary liquids in medium-size shipments. This is as expected, since the lowest rate a common carrier can quote on systematic, predictable shipments in bulk is approximately equal to cost plus a normal profit (if costs are calculated on the assumption of a high intensity of use).

[31] Furfural is subject to special consideration. See pp. 114–115.

[32] Possibly, anhydrous hydrogen chloride could be added to the third group. For the present study, it is assumed that hydrogen chloride is shipped in water solution, as hydrochloric acid. Butadiene is subject to special consideration, as noted below, p. p. 114n.

pose of determining transport cost, the weight of water must be taken into account. Commercial concentrated nitric acid contains about one-third water; commercial concentrated hydrochloric acid, about two-thirds; concentrated caustic soda solution about 73%.[33] The price differences recorded in Table 13 for the acids and for the purchase of caustic soda take the water content into account. The price difference for the sale of caustic soda is subject to special consideration and is discussed below.

The variation of transport rates on liquid chemicals with the amount shipped per year does not influence the analysis in the present study; the programs actually calculated are such that only the lowest rates (for shipments exceeding 10 MM lb/yr) are applicable.

Estimated transport costs for *solid chemicals* are based not on estimated ton-mile rates, but on published data relating to ocean freight charges between Puerto Rico and the mainland. Two categories of solid chemicals are defined: (1) bulk, impure chemicals, and (2) purified, packaged chemicals. The former are estimated to move at a total transport cost of $4600/MM lb; the latter, at $6800/MM lb.[34]

---

[33] Commercial nitric acid of the most concentrated grade (42° Bé) contains 67.2% HNO_3. Commercial hydrochloric acid of 20° Bé contains 32.0% HCl. (There also is a 22° Bé grade with 35.8% HCl content.) Caustic has typically been shipped in 50% solution; however, more recently a commercial 73% product came into use. At this concentration, caustic freezes to a solid at 45° F, and is shipped in insulated tank cars. See R. N. Shreve, *The Chemical Process Industries,* McGraw-Hill Publishing Co., 1945, pp. 408 and 415; R. A. Springer, "How to Handle Liquid Caustic," in *Manual of 224 Successful Solutions to Chemical Engineering Problems* (a collection of reprints from *Chemical Engineering*), sec. 4, p. 5 (no date).

[34] Estimated transport costs for solid chemicals are based on three references: (1) S. E. Eastman and D. Marx, *Ships and Sugar,* University of Puerto Rico Press, San Juan, 1953, pp. 162-165; (2) "United States Atlantic and Gulf-Puerto Rico Tariff, *op. cit.;* (3) *Facts for Businessmen,* Puerto Rico Economic Development Administration, pp. 65-66.

The lowest rate encountered in these sources was the one for coal in bulk, which is equivalent to $3050/MM lb. Coal in bags takes a rate of $3950/MM lb. The rate for coke in bulk is $4500. (All of these rates are from "United States Atlantic and Gulf-Puerto Rico Tariff," *op. cit.*) The lowest rate for fertilizer in bulk is $3800 (*ibid.,* Apr. 27, 1953). The rates are considerably higher for more valuable commodities. Thus, for example, the rates on high-grade fertilizer range up to $10,000 for ammonium nitrate (*ibid.,* Apr. 27, 1953). The rate on chemicals (not otherwise specified) is $14,900 (*ibid.,* Sept. 1, 1952).

The rate for bulk, impure chemicals was set at $4600/MM lb. This rate is some-

COMMODITY COST AND REVENUE DIFFERENTIALS 113

In spite of the different procedures and sources for estimation, these rates agree fairly well with those derived for liquid chemicals. For example, the $4600 transport cost estimated for bulk, impure chemicals is about 20% above that for ordinary noncorrosive liquid chemicals in large shipments. The $6800 transport cost estimated for purified, packaged chemicals falls between the estimated transport costs for medium and small shipments of ordinary noncorrosive liquid chemicals.

Commodities appearing in the programs of this study in the category of bulk, impure chemicals are lime, sulfur, coke, salt, limestone, ammonium nitrate, urea, bagasse, and ammonium sulfate. However, for the calculation of price differences, all but the first two of these are subject to special considerations to be discussed below. In the category of purified, packaged chemicals are acrylonitrile, dimethyl terephthalate, Orlon polymer, Dynel polymer, Dacron polymer, adipic acid, sodium cyanide, and nylon salt.

Transport costs are estimated for two categories of *gases*. Easily condensible gases are classed with liquids transported in pressure tanks, and are subject to the rates already given for pressurized liquid chemicals. On the other hand, some gases cannot be liquefied at reasonable pressures, and are thus usually transported in high-pressure cylinders. The transport cost on such gases is estimated to be at least $45,000/MM lb,[35] a figure that places them clearly outside realistic consideration for overseas transport for other than small-input specialty uses, such as the use of imported oxygen for welding. Among the commodities appearing in the present study, the following gases are transportable only in cylinders: hydrogen, methane, ethylene, ethane, acetylene, and nitrogen.

---

what above the one represented by coke. The rate for purified, packaged chemicals was set at the higher figure of $6800 in order to take account of the probably smaller shipments. This rate is still well below the present rates of the common carriers.

These estimates refer to the long run, and it is assumed that the rates will reflect the costs that apply when ships are used at a reasonably high intensity the year round.

[35] The transport rate on gases in cylinders, taken from the "United States Atlantic and Gulf-Puerto Rico Tariff," *op. cit.*, Nov. 3, 1952, is $14,900/MM lb, including tare. The tare is roughly estimated as about two-thirds of the total weight for ethylene, ethane, nitrogen, and acetylene; approximately five-sixths for methane; and nearly 98% for hydrogen. These proportions are based upon calculations in Vietorisz, *op. cit.*, Chap. 12.

The tare for acetylene is assumed to be approximately the same as for ethylene and ethane, even though acetylene is dissolved under pressure in acetone, rather than being compressed by itself.

Acetylene is somewhat different from the other gases in that it is dissolved in acetone, and not highly compressed by itself. Nevertheless, acetylene is transported in cylinders. Nitrogen occurs in the present study only as a by-product of negligible value and is assumed to be vented to the atmosphere; thus, it would never be transported overseas. If imported for specialty uses, it would take the high rate of gases in cylinders.

As explained above, the price difference for the *sale* of all fuel gases except acetylene is based on their Btu equivalent. Acetylene is too dangerous to burn; therefore, it is assigned a price difference based on overseas transport. This procedure tends to understate the advantage of Puerto Rico, since some acetylene could surely be sold locally for welding and other specialty uses.

COMMODITIES SUBJECT TO SPECIAL CONSIDERATION

The price differences on certain commodities are estimated on the basis of special considerations. These commodities are furfural, caustic soda, sulfur, and sulfuric acid.[36]

The price difference for *furfural* is derived in a different way from the method employed for other chemicals. The transport rate used is the usual one applicable to noncorrosive liquid chemicals. Since on the mainland both imported furfural (from the Dominican Republic)[37] and locally produced furfural (from such agricultural by-products as oat hulls and corncobs) are profitably used in nylon salt production, it seems reasonable to assume that they are competitive on a price basis. Thus, the price of furfural in the Dominican Republic may be taken to be lower than the mainland price by the amount of transport cost, namely, by

---

[36] Also it is to be noted that the price difference for butadiene should be the same as that for other easily condensible gases, i.e., $5530/MM lb in shipments exceeding 10 MM lb/yr. However, a price difference of $12,800/MM lb, based on figures published in the "United States Atlantic and Gulf–Puerto Rico Tariff," *op. cit.*, was adopted for butadiene before the amount of the transport cost on easily condensible gases was definitely established. The necessary correction of the computations was not undertaken since this would involve relatively minor adjustments; these adjustments would improve Puerto Rico's competitive position.

Since butadiene is the only easily condensible gas that appears as a net input or output (i.e., not a pure intermediate) of any selected program, no inconsistency in computations results from the adoption of the higher figure for butadiene.

[37] DuPont imports furfural based on bagasse from the Dominican Republic. See *Chemical Engineering*, **61**, 108 (Aug., 1954).

$3820/MM lb. Since Puerto Rico can also produce furfural from bagasse, as the Dominican Republic does, the price of furfural in Puerto Rico is assumed to be approximately the same as in the Dominican Republic. Therefore, relative to the mainland, Puerto Rico has an advantage of $3820/MM lb in the *purchase of furfural*, and a disadvantage of the same amount on the *sale of furfural*. Since the price difference on furfural is obtained directly from these considerations, it is not necessary to examine separately the production activity leading from bagasse to furfural; in fact, to do so would introduce an error. It is simply assumed that furfural is available in Puerto Rico in sufficient amounts at the Puerto Rico export price. The amount of furfural needed for a nylon plant of the scale assumed in this study is far less than the furfural potential of Puerto Rico based on available bagasse.[38]

The price difference on the *purchase of caustic soda* is derived in the normal way, this commodity being treated as a corrosive water solution. Special considerations apply only to the *sale of caustic soda*. This is important, since caustic appears as a joint product of chlorine electrolysis, and in general must be marketed in order to make profitable the production of chlorine. In several of the programs calculated in detail, surplus caustic is produced.

It is assumed in this study that the major consumption of caustic soda in the United States will be concentrated in the southeastern textile-producing region. Consequently, a Puerto Rico complex would ship its surplus caustic to this market; likewise, a petrochemical complex located at the U. S. Gulf Coast would also ship its surplus to this market.[39] Since this transport of surplus caustic

---

[38] The amount of sugar cane that can be grown under the existing legally fixed marketing quota is approximately 12 MM tons/yr. Assuming that 75% of this becomes green bagasse, and that 22 lb of bagasse are required per pound of furfural, the furfural potential of the island is 818 MM lb/yr; whereas the amount of furfural needed for 36.5 MM lb/yr nylon is at most 57.2 MM lb/yr. See N. Koenig, *A Comprehensive Agricultural Program for Puerto Rico*, U. S. Department of Agriculture, Washington, 1953, p. 196; and Joseph Airov, *The Location of the Synthetic Fiber Industry*, John Wiley & Sons, 1959, for references on the production of furfural from various raw materials.

[39] Viscose rayon and textile plants are large consumers of caustic. See Faith, Keyes, and Clark, *op. cit.*, p. 691. Nevertheless, there are many products—chemicals, pulp and paper, petroleum, lye and cleansers, soap, etc.—in the production of which large amounts of caustic is used. Consequently the assumption of shipment of caustic from the Gulf to the southeastern textile region must be tentative. A more detailed study of this problem was not considered justifiable, given the limited scope of this investigation.

on the mainland is unavoidable, the caustic soda revenue disadvantage of a Puerto Rico complex is considerably less than it would be if a substantial market were located at the Gulf Coast.

The export price in Puerto Rico is the price at the market minus transport cost. The transport cost is figured on the basis of an ocean shipment to Wilmington, N. C., of 1380 statute miles (1200 nautical miles), at 0.50 cent per ton-mile. The resulting cost is $4727/MM lb of 100% caustic shipped in 73% water solution.

The export price at the Gulf is the price at the market minus transport cost. This transport cost is based on a shipment by *barge*, at a rate of 0.35 cent per ton-mile, for an average distance of 750 miles.[40] The resulting cost is $1798/MM lb of 100% caustic shipped in 73% water solution.

In units of thousand dollars per million pounds, 100% caustic, and where P_{SE} represents the price in the Southeast, the price difference on the caustic sale activity between Puerto Rico and the Gulf is accordingly $(P_{PR} - P_{GC} = (P_{SE} - 4.727) - (P_{SE} - 1.798) = -2.929$.

The treatment of the price difference on the *purchase of sulfur and sulfuric acid* requires a comment. In the programs of the present study, sulfur or sulfuric acid inputs are very small. It is assumed, however, that in the future there will be a sizable local market for sulfuric acid in Puerto Rico. As with methanol, no attempt is made to estimate the extent of this market. It is simply assumed that the scale of sulfur or sulfuric acid imports or the scale of local sulfuric acid production will be sufficiently large to eliminate diseconomies of small scale in transport or in production. Accordingly, the price differences on the purchase of sulfur and sulfuric acid, respectively, are $4600 and $4250/MM lb. The former is the rate on bulk, impure, solid chemicals shipment, and the latter is the rate on large, corrosive, liquid chemicals shipment.

Sales of sulfur or sulfuric acid do not occur in the programs of the study. Should price differences for these activities be desired, these differences can be based on the general considerations discussed above in connection with price differences on solid and liquid chemicals.

STAPLE AND CONTINUOUS-FILAMENT FIBERS

The sale of staple and continuous-filament fibers is subject to special considerations because of the location of the market. As

[40] For barge rates, see Isard and Schooler, *op. cit.*, p. 20.

with caustic soda, this market is assumed to be principally in the southeastern United States, where much of the mainland textile industry is concentrated. It is postulated in this study that Puerto Rico exports its synthetic fiber product to this southeastern market.[41]

The southeastern region of the United States is the best location on the mainland not only for textile spinning, weaving, and finishing, but also for the production of synthetic fibers from their polymers. This conclusion has been established in a recent study by Airov, who evaluated the interplay of a wide range of factors, including the location of chemical raw materials, overland transport costs and rate structures, the pattern of the natural-gas pipeline network, and the availability and cost of textile labor.[42] On the basis of Airov's findings it can be assumed that the U. S. Southeast is the most relevant region with which Puerto Rico should be compared as a location for a synthetic-fiber-producing activity. Two important conclusions follow: first, the fiber-producing activity on the mainland sells its product locally (i.e., to local textile spinning, weaving, and finishing plants); second, for a mainland complex there is an unavoidable transport cost on polymers produced at the U. S. Gulf Coast (which is the best location for refinery and petrochemical operations) and shipped to the Southeast (which is the best location for synthetic fiber production). A Puerto Rico complex is not burdened with such transport, since all activities of this complex are assumed to be located close together.

In what follows, a price difference is employed for the sale of synthetic fibers which reflects the transport cost from Puerto Rico to the U. S. Southeast. This transport cost, which represents a disadvantage for Puerto Rico, is taken to be equal to the cost of overseas shipment from San Juan, P. R., to Wilmington, N. C.: namely, $18,900/MM lb.[43] At a later point, a separate correction is applied to the resulting computations in order to take account of the polymer transport cost due to split mainland location.

---

[41] This postulate is consistent with the decision, discussed in the previous chapter, to exclude textile activities from any complex to be investigated. It is taken for granted that this decision leads to considerable understatement of Puerto Rico's advantage, since textile activities are labor-intensive and labor is lower-priced in Puerto Rico than on the U. S. mainland (see below); however, in the present study the focus of interest is placed on the refinery, petrochemical, and fiber activities.

[42] Airov, *op. cit.*

[43] *Ibid.* The transport rate on package fiber, a commodity that is light but bulky, is based on volume. Thus the rate is unusually high from the standpoint of weight. The bulk density of fiber (staple or filament) is estimated at 7 cu ft/100 lb. In the

The *purchase of synthetic fiber* does not occur in the programs of the present study. The price difference for such an activity would be equivalent in absolute value to the price difference for sale, but opposite in sign, since *import* from the U. S. Southeast to Puerto Rico would be required.

FERTILIZER

The price difference on *sale of fertilizer* depends on the special assumption that Puerto Rico's imports of fertilizer can be replaced by local output valued and sold at the import price. The maximum amount of fertilizer that could be sold at the import price is equal to the island's total fertilizer consumption.[44] Since the U. S. Gulf Coast is the best *mainland* location for the production of fertilizer for Puerto Rico consumption, and currently does produce for Puerto Rico consumption, the price for fertilizer in Puerto Rico is taken to be the U. S. Gulf price plus transport cost. Such transport cost is based on the rates earlier derived for bulk, impure, solid chemicals, with one important modification. It is assumed that new production facilities for the concentrated nitrogenous fertilizers (ammonium nitrate and urea) will compete with the traditional form of nitrogenous fertilizer import, ammonium sulfate.[45] The various alternative nitrogenous fertilizers are considered equivalent when their nitrogen contents are equal. On this basis, 1 lb of ammonium nitrate is equivalent to 1.65 lb of ammonium sulfate;

"United States Atlantic and Gulf-Puerto Rico Tariff," *op. cit.*, the lowest rate on fiber is given for cotton: 31¢/cu ft (November 12, 1951). At 7 cu ft/100 lb, this figure converts to $21,700/MM lb.

It is usual to compress cotton for overseas shipment in order to reduce its bulk to the point where it takes the rates based on weight. In the tariff quoted above, the weight and volume rates are equivalent at a bulk density of 2.5 cu ft/100 lb. The information secured privately on the possibility of compressing synthetic fibers in this fashion is inconclusive. On the one hand, such a compression is judged to be technically feasible without damage to the fibers; on the other hand, fiber users are said to prefer greatly the uncompressed material. In the present study, the possibility of compression is left out of account.

[44] See Table 7, p. 65.

[45] Thus, ammonium sulfate production is not included among the activities of Table 3. However, it should be noted that ammonium sulfate may be produced as a by-product of petrochemical intermediate activity 69. It is to be noted that, according to the figures cited in Table 7, 73.2% of the total fixed nitrogen imported to Puerto Rico in 1952 came in the form of ammonium sulfate.

COMMODITY COST AND REVENUE DIFFERENTIALS 119

and 1 lb of urea is equivalent to 2.20 lb of ammonium sulfate.[46] Accordingly, the transport cost which determines the price difference is the transport cost of the stated amounts of ammonium sulfate per pound of ammonium nitrate or urea. The price differences so obtained are: ($4600)(1.65) = $7590/MM lb ammonium nitrate; ($4600)(2.20) = $10,120/MM lb urea.[47]

It is to be emphasized that these price differences apply only to the local fertilizer output which replaces imports from the Gulf Coast. Should local production of fertilizer exceed local consumption, the excess would have to be exported to the U. S. mainland. The export price in Puerto Rico would be $4600 below the Gulf price. Moreover, if fertilizers were exported at this price, it would not be possible to maintain the price level for local sales at the import price. Since in the present study fertilizer production is restricted to estimated local demand, these considerations are not relevant.

Fertilizer purchases by the complex do not occur in the present study, since fertilizers do not form an input for any activity within the complex. If a price difference for fertilizer purchases is desired, it is of course the same as the price difference for import-replacing fertilizer sales.

MISCELLANEOUS (ZERO-DIFFERENTIAL) COMMODITIES

In the present study, air, water, salt, limestone, carbon dioxide, and methanol are assigned a zero price difference. Air is an obvious ubiquity available everywhere at zero cost; it is a chemical input of ammonia and ethylene oxide production among the activities considered. Water is a less obvious ubiquity; in fact, industrial processes which use large amounts of water are strongly attracted to locations where water of the required quality is plentiful. However, although the locational pull of water is often decisive in

---

[46] These data were derived from the appropriate chemical formulas, impurities being neglected.

[47] It is to be noted that in the "United States Atlantic and Gulf–Puerto Rico Tariff," *op. cit.*, the rate on heavy nonconcentrated fertilizers (acid phosphate, ammonium phosphate, sulfate, phosphate rock) shipped in bulk is about $3800/MM lb. In contrast, the rate on ammonium nitrate is $10,000/MM lb (Apr. 27, 1953). Thus, the current shipping-rate structure puts the shipments of concentrated fertilizer at a considerable disadvantage on a nitrogen-equivalent basis.

selecting plant sites within a given region, it is rarely a factor in interregional comparisons, since most regions have some sites where the required water is available. Moreover, when water is really scarce, the consumption of water can be drastically reduced by treatment and recirculation, although such activities definitely increase costs. In the present study, water is considered to be a *regional* ubiquity; i.e., it is considered to be available at approximately equal cost in a few favorable locations in each region. The same assumption is made for salt and for limestone; it is simply posited that they can be produced at virtually equal cost from natural resources occurring at some locations within each region.

Finally, carbon dioxide and methanol are also assigned a zero price difference, even though these commodities are produced by distinct processes requiring inputs of fuel, power, labor, etc. Carbon dioxide is produced from flue gases or from lime-kiln gases by a very simple absorption process requiring only minor process inputs; therefore, any regional differences in production costs are likely to be negligible. Methanol, on the other hand is assumed to be produced in Puerto Rico jointly with ammonia, since the processes are closely related; therefore, a price difference for methanol ought logically to be derived. However, we do not derive such price difference since the inputs of methanol to programs of the present study are always so small that any resulting differences in production cost may be considered negligible.

Air and water inputs to various activities are omitted from Table 3; however, the inputs of the other commodities with a zero price difference are included. If the reader desires to associate a nonzero price difference with the purchase of any of these commodities, the listed inputs permit the calculation of the necessary corrections.

LABOR

Price differences are estimated separately for labor of textile grade (unskilled or semiskilled) and of chemical–petroleum grade (skilled). The estimates are adapted principally from the results of detailed studies by Airov.

The average U. S. wage rate for synthetic fiber production in January 1954 was $1.79 per man-hour.[48] On the basis of a special

---

[48] Airov, *op. cit.*, cited from *Monthly Labor Review*, **77**, 706 (1954).

COMMODITY COST AND REVENUE DIFFERENTIALS 121

regional tabulation obtained from the U. S. Bureau of Labor Statistics, Airov assumes that essentially the same wage rate ($1.78) is applicable for new synthetic fiber plants in the South. For Puerto Rico, Airov employs alternative rates of $1.25, $1.00, and $0.75 per hour. There is no basis for a more accurate estimate, since wage rates in Puerto Rico are legally subject to adjustment on the basis of the profitability of an enterprise. However, in this study, a single price difference, $0.75 per man-hour, for textile grade labor is adopted.

For chemical-petroleum labor, the wage rate for the Gulf Coast region as of January 1954 is estimated by Airov at $2.22 per man-hour. This figure is based on Bureau of Labor Statistics regional comparisons for the years 1946, 1949, and 1951, and on national average figures for January 1954.[49] For Puerto Rico, Airov assumes that the cost of direct chemical labor would be at least one and two-thirds of the U. S. average of $2.06, i.e., $3.42 per man-hour. Airov's assumption is based on the premise that skilled workers would have to be brought to Puerto Rico from the U. S. mainland. For the present study, this premise is accepted. However, considering the long-run nature of the study and the fact that more skilled labor is being trained locally in Puerto Rico, a $1.00 per man-hour price difference is judged to be adequate. Moreover, the results based on this price difference are compared (in Chapter 6) with results obtained on the assumption that the price difference on skilled labor is entirely eliminated.

Like all other price differences, those for skilled labor may be changed to suit different interpretations of the empirical data, without affecting the structure of the analysis.

ADJUSTMENTS FOR DIFFERENT RAW-MATERIAL SOURCES

In the remaining part of this chapter are discussed several adjustments, or corrections. In one sense these adjustments are differential profitability corrections and as such logically ought to be included in later chapters. However, these corrections may also be

---

[49] References cited by Airov are: (1) "Wage Structure, Industrial Chemicals," U. S. Bureau of Labor Statistics: series 2, no. 25, p. 14, Table 4 (1946); series 2, no. 73, p. 4, Table 2 (1949); series 2, no. 87, Table 4, p. 7 (1951); and (2) *Monthly Labor Review*, **77**, 471-2 (1954).

viewed as the result of price differences; and, since they can thus be conveniently treated at this point, we have chosen to do this.

The *production of ammonia from fuel oil* in Puerto Rico requires one such correction, since a complex located at the U. S. Gulf Coast would always produce ammonia from fuel gas. Fuel gas is 14.5¢/MM Btu cheaper than fuel oil at the Gulf; and, as already discussed, all fuel oil produced by a complex at the Gulf would be exported to New York. On the other hand, in Puerto Rico, all fuels are assumed to have a price equal to the fuel oil export price.

The disadvantage of producing ammonia from fuel oil in Puerto Rico is twofold. First, a more expensive fuel is used, compared with that on the Gulf Coast. Second, more of it is used, since ammonia production from fuel oil requires a slightly higher Btu input than ammonia production from fuel gas. In the present study, the latter effect is neglected; and, in consequence, a price difference of 14.5¢/MM Btu is charged against all fuel oil used for ammonia production in Puerto Rico. On this basis, a disadvantage for Puerto Rico of $20,900 per unit level of activities 37 and 38 is obtained.[50]

ADJUSTMENTS BECAUSE OF SPLIT-LOCATION PATTERN ON MAINLAND

As already indicated, an efficient mainland complex would involve a split-location pattern which would require the transport of polymer from the Gulf Coast to the southeastern textile region.[51] The amount of the polymer transport cost constitutes a correction which must be added to the advantage of a Puerto Rico complex.

---

[50] An exact treatment of the problem would consist of two steps: (1) A price difference of 14.5¢/MM Btu would be applied to all fuel gas replacing fuel oil at the Gulf in its use as a raw material for ammonia; (2) the full cost of the extra Btu needed in Puerto Rico as a result of using fuel oil rather than fuel gas would be charged against the advantage of the complex in Puerto Rico. For the latter purpose, the absolute price of fuel oil in Puerto Rico would be required.

The simple (approximate) correction adopted in this study is derived as follows: At unit scale, activities 37 and 38 take an input of 0.023 MM bbl of fuel oil (cycle oil or heavy residual); the corresponding Btu at 6.25 MM Btu/bbl is $0.023 \times 10^6 \times 6.25 \times 10^6 = 1.44 \times 10^{11}$ Btu. At 14.5¢/MM Btu, this yields a correction of $14.5 \times 10^{-2} \times 10^{-6} \times 1.44 \times 10^{11} = \$20,900$ per unit scale of activities 37 and 38, to be *deducted* from the advantage of a complex in Puerto Rico.

[51] In the case of nylon fiber production, nylon salt is transported. Therefore, in this connection, nylon salt is included in the "polymer" category.

COMMODITY COST AND REVENUE DIFFERENTIALS

The correction is calculated on the basis of transport cost from Orange, Tex. to Chattanooga, Tenn. Airov has estimated this cost as $8800/MM lb. The adoption of this figure in the present study involves the assumption that the cost of fiber shipment by truck or by rail from Chattanooga, Tenn., to actual consumption points is equal to the cost of similar shipments from Wilmington, N. C. The latter shipments would be required for the fiber output of the Puerto Rico complex.

The correction is applied to the fiber-production activities and has a value of $88,000 per unit level of any fiber-production activity (i.e. activity 47, 48, 50, 51, 53, 54, 72, or 73), since these activities each take a polymer input of 10 MM lb at unit level.

A further correction is made necessary from the fact that the cost of fuel, steam, and power inputs in the textile area of the southeastern United States is different from the cost of these inputs at the Gulf Coast. On the premise that the fuel will be natural gas transported by pipeline, the price at the fiber location will be the Gulf price for fuel gas plus 9.8¢/MM Btu.[52] Accordingly, the disadvantage of Puerto Rico on fuel account is overstated by 9.8¢/MM Btu for that portion of the total fuel input that is incurred in connection with fiber production. The necessary correction, added to the advantage of the complex in Puerto Rico, is the following: For staple fiber activities 47, 50, 53 operated at unit scale (10 MM lb/yr output), $78,900; for nylon staple activity 72 at unit scale, $80,080; for continuous-filament activities 48, 51, 54 at unit scale, $87,600; for nylon continuous filament activity 73 at unit scale, $88,780.[53]

It should be noted that, if fuel, steam, and power inputs to the fiber activities had been defined as distinct commodities, a price

---

[52] The transport cost of gas by pipeline is assumed to be 1.4¢/M cu ft per 100 miles; assumed heating value of gas is 1000 Btu/cu ft; assumed distance to southeastern textile location is 700 miles.

[53] The Btu requirement for steam and power was figured at 1500 Btu/lb and 4500 Btu/kw-hr, respectively. Accordingly, the corrections were calculated as follows:

Nonnylon staple, $(0.500 \times 10^9 \times 1500 + 12.000 \times 10^6 \times 4500)9.8(10^{-2})(10^{-6})$
Nylon staple, $(0.505 \times 10^9 \times 1500 + 13.000 \times 10^6 \times 4500)9.8(10^{-2})(10^{-6})$
Nonnylon filament, $(0.550 \times 10^9 \times 1500 + 15.000 \times 10^6 \times 4500)9.8(10^{-2})(10^{-6})$
Nylon filament, $(0.555 \times 10^9 \times 1500 + 16.000 \times 10^6 \times 4500)9.8(10^{-2})(10^{-6})$

The amount of by-product power available at unit scale for staple fiber production is about $(0.500)(10^9)(60,000)(10^{-6}) = 30.00$ MM kw-hr. These amounts exceed considerably the required inputs for unit scale, 12.00 MM kw-hr and 15.00 MM kw-hr. For references regarding the constants employed in these calculations, see footnotes 13, 15, and 16.

difference could have been associated with each of these commodities and the above correction would have been unnecessary. The price difference for fuel inputs to fiber activities would have been $19.9 - 9.8 = 10.1$¢/MM Btu; the price differences for steam and power could then have been derived from this figure in the usual manner.

Chapter 6

Calculation of Preliminary Net Advantages for Puerto Rico Complexes

In Chapter 4, the amounts of net inputs and outputs associated with the eighteen selected Orlon, Dynel, and Dacron programs were calculated. In Chapter 5, the commodity-price differences applicable for comparison of a Puerto Rico and a mainland operation were derived. The present chapter describes the manner in which these two sets of data are combined to yield, for each of the eighteen programs, locational advantage or disadvantage in Puerto Rico as compared with an essentially identical program on the mainland. In addition, this chapter contains a description of a short-cut method for calculating the locational advantage or disadvantage term—a method that by-passes the computation of total net inputs and outputs of the program under consideration. This short-cut method is applied to the ten selected nylon programs.

In the last part of this chapter, the results shown by these provisional calculations of locational advantage and disadvantage are evaluated. To aid the evaluation, a number of possible alternative programs are compared in this chapter to the twenty-eight that have been developed in detail. These alternative programs, with respect to a Puerto Rico location, all require the import of some of the intermediate commodities, and thus do not provide for the full

126 *INDUSTRIAL COMPLEX ANALYSIS*

range of activities found in the original twenty-eight programs. They are called "short programs."[1]

ELEMENTS OF LOCATIONAL ADVANTAGE: THE FULL PROGRAMS

The computations associated with the Orlon A program may be used to illustrate the general method for calculating a given program's locational advantage or disadvantage in Puerto Rico.

To start, consider the net inputs and outputs of the Orlon A program recorded in column 1 of Table 11. First listed is the input of crude oil (-9.428 MM bbl). The price difference for crude-oil *purchase* ($143,000/MM bbl) appears in Table 13, part 1, column 1. Multiplication of these two figures yields $-\$1.348$ MM. The negative sign indicates that this figure is Puerto Rico's *disadvantage* relative to the mainland on crude-oil account for the Orlon A program. In contrast, the second listed item in column 1 of Table 11 is the output of straight-run gasoline ($+1.300$ MM bbl). The price difference for the *sale* of gasoline ($73,000/MM bbl) is recorded in Table 13, part 1, column 2. Multiplied together, these two figures yield a figure of $0.095 MM. The fact that this figure is positive indicates that it is an *advantage* for Puerto Rico relative to the mainland.[2]

In a similar fashion there can be calculated the cost or revenue advantage or disadvantage for Puerto Rico for each of the other Orlon A inputs and outputs listed in Table 11. Summing all these advantages and disadvantages yields the net advantage or disadvantage for Puerto Rico compared to the mainland with reference to those inputs and outputs listed in Table 11. The net figure, which is recorded in column 1, Table 15, amounts to $-\$1.702$ MM. It is a disadvantage for Puerto Rico.

---

[1] In addition to the full and short programs examined in this chapter, one other program, to be termed a reduced Dacron A program, will be found to have significant advantages—but only after the analysis of the differential profitability corrections in the succeeding chapters has been completed.

[2] Since in Table 11 all net inputs have a negative sign and all net outputs a positive sign, and since the price differences of Table 13 all refer to the Puerto Rico price *minus* the mainland price, the product of the multiplication of any item in Table 11 by the corresponding price difference in Table 13 represents an advantage for Puerto Rico if the resulting sign is positive, a disadvantage if negative.

TABLE 15
PRELIMINARY LOCATION ADVANTAGES OF SELECTED FULL COMPLEXES IN PUERTO RICO
(IN $M/YR)

Complex		Table 3 Inputs and Outputs	Chemical Petroleum Labor	Textile Labor	Mainland Polymer Transport	Tennessee-Texas Fuel Cost Difference	Ammonia Process Difference	Net Total, cols. 1-6
		Advantages Due to:						
		(1)	(2)	(3)	(4)	(5)	(6)	(7)
Orlon	A	-1702	-2296	3978	321	288		589
	B	-1218	-2329	3986	321	288	-72	976
	C	-1514	-2275	3973	321	288		793
	D	-1294	-2305	3980	321	288	-33	957
	E	-1592	-2580	4044	321	288		481
	F	-1422	-2591	4047	321	288	-25	618
	H	-1738	-2179	3951	321	288		643
	J	-1174	-2217	3960	321	288	-85	1093
Dynel	A	-1369	-2590	4046	321	288		696
	B	-1290	-2431	4010	321	288		898
	C	-1285	-2383	3998	321	288		939
	D	-1139	-2631	4056	321	288		895
	E	-1131	-2463	4017	321	288		1032
	F	-1046	-2390	4000	321	288		1173
Dacron	A	-872	-2229	3963	321	288		1471
	B	-920	-2473	4019	321	288		1235
	C	-948	-2226	3962	321	288		1397
	D	-997	-2469	4018	321	288		1161
Nylon	A	-1070	-2624	4055	321	292		974
	B	-1045	-2658	4063	321	292		973
	C	-1147	-2724	4078	321	292		820
	D	-1417	-2519	4031	321	292		708
	E	-1391	-2553	4038	321	292		707
	F	-1493	-2619	4053	321	292		554
	G	-1385	-2275	3974	321	292		927
	H	-1320	-2321	3985	321	292		957
	J	-925	-2665	4064	321	292		1087
	K	-1622	-2541	4035	321	292		485

Substitution of continuous filament for staple fiber activities adds $2,704M/yr to each net advantage figure.

For the remaining Orlon programs and for the Dynel and Dacron programs, similar computations may be performed. The resulting net figures are recorded in the first column of Table 15. The amounts are all negative; they represent net disadvantages for Puerto Rico relative to the mainland with reference to the inputs and outputs recorded in Table 11.

Also in the first column of Table 15 are net figures relating to nylon programs. These figures were calculated in a different manner. The computation of total net inputs and outputs was avoided. Instead, for each activity of Table 3, a calculation was made of Puerto Rico's advantage or disadvantage relative to the mainland in operating the activity *at unit level*. For any given activity, the calculation consisted of: (1) multiplying each input and output appearing in the relevant column of Table 3 by the appropriate price difference, and (2) summing the resulting products.

Once the unit scale advantage or disadvantage was established for each activity of Table 3, the net advantage or disadvantage of Puerto Rico, on account of the input and output commodities listed in Table 3, was determined for each program. For each program the two steps involved were: (1) multiplying the scale (in multiples of unit level) of each activity required by the program by that activity's unit scale advantage or disadvantage, and (2) summing the resulting products.

As indicated above, such computations were performed for the ten nylon programs. The derived amounts, all representing net disadvantages for Puerto Rico, appear in the first column of Table 15. The same amounts, of course, would have been obtained had the method of calculation used for the Orlon, Dynel, and Dacron programs been employed.

Elimination of the computation of aggregate inputs and outputs of individual commodities for each separate program effects a very considerable saving of time. However, it must be emphasized that the use of a single set of unit-level advantages or disadvantages for the activities of Table 3 requires that, in all the programs for which calculations are made, the price of each input and output commodity is either always the import price or always the export price. This requirement is met in all the nylon programs of this study. However, suppose that such were not the case. For example, suppose that the operation of a program, say program X, results in the production of fuel oil in excess of the program's internal fuel requirements, but that the operation of a second program, say program Y, results in an overall fuel deficit. For program X the relevant price for fuel oil would be the *export* price, whereas for program Y it would be the *import* price. Since the price difference applied to fuel inputs generally is dependent upon which fuel oil price is relevant, the unit-scale advantage or disadvantage of any program-X activity requiring a fuel input would differ from that of the same activity in program Y. In such a situation the "short-cut" method of calculating net advantage or disadvantage would be inapplicable, at least with regard to the use of a single set of unit-level advantage or disadvantage terms for the activities common to both programs.[3]

Once the advantage or disadvantage of Puerto Rico for any complex is determined for all commodities listed in the matrix of Table 3, it becomes necessary to account for the advantages or disadvantages associated with labor inputs, inputs which, as already noted, do not vary linearly with scale.[4] The magnitude of these advantages or disadvantages is derived in two steps. The first step involves the multiplication of the total inputs of "imported" labor (direct and indirect labor of the chemical–petroleum type, as presented in Table 12) by the relevant price difference of $1.00 per man-hour. (See the discussion in Chapter 5.) The results of this first step are recorded in column 2 of Table 15. As is to be expected, all the figures are negative, indicating a disadvantage for Puerto Rico complexes.

The second step involves the multiplication of the total inputs of "local" labor (direct and indirect labor of the textile type, as presented in Table 12) by the relevant price difference of $0.75 per man-hour. (See the discussion in Chapter 5.) The results of this second step are recorded in column 3 of Table 15. As is to be expected, all the figures are positive, indicating an advantage for Puerto Rico complexes.

[3] When large numbers of programs must be evaluated, it is often possible to divide them into two or more subgroups. For each program in the same subgroup the same sale–purchase price pattern for commodities would obtain in the application of the short-cut method. Usually the particular subgroup to which a given program belongs can be determined by noting whether one or two key commodities (e.g., fuel oil) show a deficit or a surplus in the program.

A possible objection to the use of the short-cut method is that it provides no check on errors committed in setting up the programs. Common types of errors include the omission of activities, over- or underprogramming of intermediate commodities, use of erroneous commodity units, and the application of improperly derived price differences on minor by-products or input commodities. When the longer, general method of calculation is employed, many of these errors show up clearly as inconsistencies in the data of the table of net inputs and outputs (e.g., Table 11).

To reduce the danger of such errors and yet secure the savings of the short-cut method, one or two programs of each subgroup may be calculated by the general method and the net inputs and outputs examined. In this manner at least the systematic errors may be caught and corrected. Then the remaining programs can be calculated by the short-cut method.

[4] Because of the nonlinearity of labor inputs, the labor cost advantage or disadvantage cannot be incorporated in an activity's unit-level advantage or disadvantage. Hence both the general and the short-cut methods must include the same side calculations on labor.

Additionally, an advantage accrues to Puerto Rico complexes because they can be spatially juxtaposed and avoid interregional transport of polymer. The amount of the advantage is the same for every program. It is $321,000/yr, and is computed by multiplying the unit-scale advantage for any fiber-production activity, namely, $88,000 (as derived in Chapter 5), by the output, in multiples of unit level, of the fiber activity, namely 3.65 (which is the same for every program).

As discussed in Chapter 5, the disadvantage of Puerto Rico complexes on fuel inputs for fiber-production activities is overstated.[5] The amounts representing the necessary correction are given in column 5 of Table 15. For all Orlon, Dacron, and Dynel programs, the correction is the same, namely $288,000/yr, and is computed by the multiplication of the unit-scale correction for these fiber activities, namely, $78,900 (as derived in Chapter 5) by the output, in multiples of unit level, of the activities, namely 3.65. For each of the nylon programs the correction amounts to $292,000/yr, computed by multiplication of $80,080 (the amount of the unit-scale correction) by 3.65.

Still another adjustment is required to correct the understatement of the disadvantages of Puerto Rico programs involving the production of ammonia from fuel oil. The correction reflects the fact that a mainland (Gulf Coast) complex would always manufacture ammonia from natural gas instead of the more expensive material, fuel oil. As explained in Chapter 5, the amount of the correction for unit-level operation of the ammonia-from-fuel-oil activities is $20,900. The four programs that require the production of ammonia from fuel oil are Orlon B, Orlon D, Orlon F, and Orlon J. The scale (in multiples of unit level) of the relevant ammonia activity (38) for these four programs are 3.461, 1.569, 1.207, and 4.067, respectively.[6] Multiplication of the unit-level correction by these activity scales yields, respectively, the required adjustments. They are recorded in column 6 of Table 15.

The results of the summation by program of the data in the first six columns of Table 15 are given in column 7. The resulting positive amounts indicate that for each of the twenty-eight selected programs there is a net locational advantage of a Puerto Rico

---

[5] This overstatement results from application of the same price difference to all fuel inputs, whereas, because of the split location of mainland complexes, the amount of the price difference on fuel used in fiber activities actually should be somewhat less.

[6] These scales are recorded in the relevant individual Orlon-production programs in Appendix B.

PRELIMINARY NET ADVANTAGES 131

complex over an essentially identical, efficiently located mainland complex.[7] This quantitative locational comparison—which has been the object of the first major analytical stage of this study—may be viewed as a preliminary approximation. Questions of the "differential-profitability" corrections necessary to take account of the evident nonidentical aspects of Puerto Rico and mainland complexes are considered in ensuing chapters.

It was pointed out in Chapter 4 that a principal objective in the identification of specific programs for analysis was to choose a group of programs representative of the major processing chains leading from crude oil through petrochemical intermediates to synthetic fibers and fertilizers. The accomplishment of this objective within the set of general limitations and constraints adopted for the programs resulted in some instances in violations of the minimum-scale output capacities recorded in Table 8. Such violation means that certain activities are scheduled at scales that are smaller than those generally considered economically feasible in the United States. It must be borne in mind that the activities are *technically* feasible at the small scales required. The only justifiable inference to be drawn from the minimum-scale violations is that the ultimate locational advantage or disadvantage for Puerto Rico of the programs in which they occur is likely to be substantially influenced by comparative economies of scale.[8]

The effects of scale economies on the overall profitability of each of the selected programs are incorporated in the differential profitability calculations of subsequent chapters. However, it is of interest at this juncture to note that substantial violations of minimum-scale restrictions affect only four activities, i.e., the production of hydrogen cyanide (in Dynel A, Dynel B, and Dynel D), ethylene (in Dynel D, Dynel C, Dacron A, Dacron B, and Dacron C), sulfuric acid (in Dacron C and Dacron D), and benzene (in Nylon C and Nylon F). Furthermore, there is reason to believe

---

[7] For four of the Orlon programs, the Puerto Rico and mainland complexes are not strictly identical, since the former include production of ammonia from fuel oil and the latter do not. As indicated by the amounts in column 6, the resulting corrections to the overall advantage figures are relatively minor.

[8] It was noted in Chapter 4 that in three programs the *maximum*-scale limitation on refinery operation is exceeded. Also in four of the Orlon programs the scheduling of fertilizer activities deviates from the "standard" or upper-scale limit. These latter programs restrict the production of ammonia to the use of available refinery-gas materials instead of bringing ammonia production up to "standard" fertilizer requirements by utilizing fuel oil as a raw material.

that, for most of these commodities, total local demand in Puerto Rico and the Caribbean area will be sufficient to warrant production on considerably larger scales than those specified solely for the programs of this study. For example, sulfuric acid is a basic chemical with diverse industrial uses. Consequently a sulfuric acid plant of well above minimum economic size could doubtless be included in the Puerto Rico complex.[9] Also although ethylene and hydrogen cyanide do not have the range and diversity of uses that sulfuric acid does, they both have important applications other than those considered for the programs of this study. Hence, it is possible that a Puerto Rico complex could include ethylene and hydrogen cyanide activities of at least minimum economic scale.[10]

It should be emphasized that the operation of certain activities above the scale required in the selected programs of this study would not necessarily constitute an unqualified improvement in the profitability of a Puerto Rico complex. The activities may be subject to disadvantages other than those of small scale. Indeed, it is conceivable that, on balance, increased scales of operation could be more disadvantageous than the small scales. A general evaluation of the desirability of increasing the scales of petrochemical activities in the Puerto Rico complexes is presented in the concluding chapter of this study.

ELEMENTS OF LOCATIONAL ADVANTAGE: THE SHORT PROGRAMS

As the reader may recall, the decision was made to limit detailed analysis to programs representative of a full complex of refinery–petrochemical–fertilizer–synthetic fiber activities. (See the discussion of Chapter 2.) However, alternative types of programs may be feasible and indeed favorable for Puerto Rico. These alternative programs can be constructed by substituting imported commodities for various intermediate activities of the full programs, or by changing the number or type of output commodities.

---

[9] The favorability of such a plant would be further enhanced if it could utilize the by-product sulfur recovered from desulfurization of sour Venezuelan crudes fed to the refinery.

[10] Subsequent differential profitability calculations indicate that the magnitude of scale economies in benzene production is very great. As a consequence, even though more than one use for benzene may be anticipated in Puerto Rico, it is most realistic to assume that benzene would be *imported* from the mainland.

It is therefore interesting and significant to compare the results obtained from the computations for the full programs with the results for several selected *short programs,* each of which involves only a few activities. If the cost–revenue differential calculations for any of the short programs should show a greater net advantage for Puerto Rico than those for the full programs, the amount of the excess advantage furnishes some indication of the extent to which economies of spatial integration and the nonquantifiable economic and social advantages of *full* complexes must exist if full complexes are to be the most desirable ones for Puerto Rico.

At this point, the short programs chosen for analysis are described and the reasons for their selection set forth. Then the results of the locational comparison of the short programs in Puerto Rico with identical mainland programs are summarized. It is evident that the final comparison of the short programs with the full programs must wait for the calculation of differential profitability corrections for both types. However, the inclusion here of preliminary short program results permits the analyst later to note differences between the short programs and the full programs resulting from differential profitability corrections alone.

As a first step in the process of selecting favorable short programs, Table 15 is examined. Such examination indicates that the net advantage of a Puerto Rico location for any of the full programs is primarily due to the magnitude of Puerto Rico's advantage in the cost of local (textile-type) labor. Local labor inputs comprise not only the direct labor required for the fiber activities but also portions of the indirect labor inputs for all activities. However, none of the nonfiber activities has an input of *direct* local labor. Hence, for all nonfiber activities the disadvantage on imported (chemical–petroleum) labor outweighs by far the advantage on local labor.[11] With respect to labor inputs, then, the fiber activities are the only ones that make a positive contribution to locational advantage.

As a consequence, one type of short program which merits consideration involves the production of fiber by itself. *Short Program A* thus calls for the annual production and export to the mainland of 36.5 MM lb of staple fiber from Orlon, Dynel, or Dacron polymer, or nylon salt imported from the Gulf Coast. In the absence of an

---

[11] For nonfiber activities, the direct labor inputs, which are all subject to a disadvantage of $1.00 per man-hour, are always greater than the indirect labor inputs, only part of which are subject to an advantage of $0.75 per man-hour.

oil refinery, fuel oil would have to be imported to Puerto Rico to provide for the steam and power requirements of the fiber activity. Based on the fuel-oil import price, the net advantage of Short Program A in Puerto Rico over an identical mainland operation is $1.563 MM/yr, as recorded in column 2, Table 16.[12]

It is possible that a refinery conducted independently of the fiber operation might supply fuel oil at the fuel-oil export price. Under such conditions Short Program A in Puerto Rico would enjoy a net advantage of $1.595 MM/yr. (See column 1, Table 16.)

The figures of Table 14 indicate that all the programs suffer net disadvantages if the influence of labor inputs is not considered. However, closer examination of the figures for the Orlon programs shows that, other things being equal, programs including the maximum production of fertilizer have smaller net disadvantages on nonlabor inputs and outputs than programs with less than the maximum fertilizer output.[13] This fact suggests the consideration of fertilizer activities alone. *Short Program B* accordingly involves (*a*) the production of 80 MM lb/yr of ammonia from fuel oil, and (*b*) the use of the ammonia as raw material for the production of the intermediate, nitric acid, and the fertilizer commodities, urea and ammonium nitrate.

The use of price differences based on the fuel-oil import price yields a net advantage figure for Short Program B in Puerto Rico of $724,000/yr compared with an essentially identical mainland program.[14] If use of the fuel-oil export price were warranted, the same program would yield a net advantage for Puerto Rico of $771,000/yr. (See columns 1 and 2, Table 16.)

In addition to short programs involving fiber activities alone and fertilizer activities alone, it is of interest to consider a program involving an oil refinery by itself. Accordingly, *Short Program C* calls for the operation of a refinery (prototype 4) without any asso-

---

[12] This figure is calculated in accordance with the general method described in this chapter. Relevant price differences are the ones developed in Chapter 5.

All staple fibers except nylon yield the same result for Short Program A since their inputs and outputs (Table 3) are essentially the same from the standpoint of computing cost and revenue differentials. The nylon staple activity has slightly greater steam and power inputs, but the difference in final results would be insignificant—of the order of $2000/yr.

[13] Orlon programs A, C, E, and H are the same as Orlon programs B, D, F, and J, respectively, except that the former call for less than the maximum fertilizer production.

[14] This figure results from computations that include the correction for mainland use of cheaper natural gas as raw material for ammonia production.

TABLE 16
PRELIMINARY LOCATIONAL ADVANTAGES OF SELECTED SHORT PROGRAMS IN PUERTO RICO

Short Program	Description	Based on Fuel Oil Export Price (1)	Based on Fuel Oil Import Price (2)	Based on Fuel Oil Export Price and		
				Usual Price Differences (3)	No Disadvantage on Short Haul (4)	No Disadvantage on Short Haul and 2¢/b Disc. on Crude (5)
A	Staple Fiber Alone	+1595	+1563			
B	Fertilizer Based on Fuel Oil	+771	+724			
C	Refinery #4 Alone			-704	-300	-112
D	Fiber and Fertilizer	+2366	+2287			
E	Refinery #4 and Fertilizer			+67	+471	+659
F	Refinery #4 and Fertilizer and Staple Fiber			+1662	+2066	+2254

Notes:

Units for Locational Advantage, $M/yr.

ciated fertilizer or fiber activities. In order to encompass a wider range of possibilities for comparison among the short programs and between the short and the full programs, the calculations with respect to Short Program C have been performed on the basis of three separate sets of price differences. First to be employed were the usual price differences on liquid commodities and on gaseous fuel products. These have been developed in Chapter 5 and summarized in Table 13. On this basis, unit-level operation of the refinery by itself in Puerto Rico would be subject to a disadvantage of $704,000/yr when compared to an identical mainland operation. (See column 3, Table 16.)

The second set of price differences was based on the assumption that the Puerto Rico refinery would be subject to no short-haul disadvantage in ocean transport cost.[15] Under such conditions, the disadvantage of the Puerto Rico refinery would be $300,000/yr. (See column 4, Table 16.)

The third set of price differences employed is exactly the same as the second set, except for the price difference on crude oil. For this set of computations, Venezuelan crude is taken to be subject to a price discount of 2¢/bbl in New York relative to the delivered price of Gulf Coast crude. (See the discussion of this assumption in Chapter 5.) Such a market situation would definitely enhance the relative profitability of the Puerto Rico refinery. Computation shows that its net disadvantage compared with an identical mainland operation would be only $112,000/yr. (See column 5, Table 16.)

The remaining three short programs selected for evaluation consist of combinations of the three short programs already described. *Short Program D,* comprising fiber and fertilizer activities, is a combination of Short Programs A and B. Combining the activities does not change any of the relevant price differences. Therefore, the preliminary net advantage of Short Program D in Puerto Rico is obtained merely by adding the relevant preliminary net advantages of Short Programs A and B. The resulting amount is $2.287 MM/yr if the required fuel oil for fuel inputs and ammonia production must be imported. If fuel oil could be obtained locally

---

[15] Specifically, the price differences were derived on the assumption of ocean transport by tanker at a rate of 1.8 cents per 100 nautical barrel miles, regardless of the distance shipped. (Compare the assumptions adopted in Chapter 5 for rates used in the general computations.)

As mentioned previously, the cost disadvantage of short runs can largely be eliminated by the construction of ships and terminals designed for unusually rapid loading and unloading.

PRELIMINARY NET ADVANTAGES 137

at the fuel-oil export price, Short Program D's preliminary net advantage in Puerto Rico would be $2.366 MM/yr. (See columns 1 and 2, Table 16.)

Short Program E comprises a combination of refinery and fertilizer activities. The amount of the program's preliminary net advantage in Puerto Rico is obtained by summing the preliminary net advantage of Short Program B (fertilizer) and the preliminary net disadvantage of Short Program C (refinery). Since the refinery would produce fuel oil for export, the relevant amount for the advantage of the fertilizer activities is the amount calculated on the basis of the fuel-oil export price. Furthermore, since there are three separate figures for the preliminary disadvantage of the refinery activities, corresponding to the three sets of assumptions regarding transport costs on crude oil and refined products, there are accordingly three separate figures for the preliminary net advantage of Short Program E. This advantage is $67,000/yr if the transport rates generally postulated in the study are employed. It is $471,000/yr if no short-haul transport-cost disadvantage is incurred. It is $659,000/yr if the further postulate of 2¢/bbl discount on Venezuelan crude is adopted.[16] (See columns 3, 4, and 5, Table 16.)

The third of the three combination short programs contains refinery, fertilizer, and staple fiber activities. It is designated *Short Program F*. The preliminary net advantage of this short program in Puerto Rico is obtained by adding the preliminary advantage term for Short Program A (based on the fuel-oil export price) to the preliminary net advantage term for Short Program E (refinery and fertilizer activities).

With price differences based on the originally postulated ocean

---

[16] It might appear incorrect to add the amounts of Short Programs B and C, since the fertilizer production of Program B is based on fuel oil whereas in the combined Program E the fertilizer activities would be based on refinery gas materials. However, the calculation of the preliminary net advantage of Program B includes the standard differential profitability correction on ammonia production from fuel oil. As a result of this correction, the amount of Program B's advantage is the same as it would be if the fertilizer activities were based on gaseous raw materials.

The preliminary net advantage terms for Program E under the two special sets of assumptions regarding ocean transport costs of crude oil and refined products are slightly overstated. The reason is that, under these assumptions, the disadvantage of Puerto Rico on account of fuel inputs is slightly greater than it is on the basis of the original set of transport costs. Thus the net advantage on fertilizer activities would be slightly less than was calculated for Program B. The error is very minor. It is approximately $3000/yr for Short Program E.

138   INDUSTRIAL COMPLEX ANALYSIS

transport rates, the preliminary net advantage for Short Program F in Puerto Rico is $1.662 MM/yr. If the assumption of no short-haul transport-cost disadvantage is adopted, this program's preliminary net advantage is $2.066 MM/yr. If the further assumption of a 2¢/bbl discount on Venezuelan crude is accepted, Short Program F shows a preliminary net advantage in Puerto Rico of $2.250 MM/yr.[17] (See columns 3, 4, and 5 of Table 16.)

PRELIMINARY EVALUATION OF PROGRAMS

The figures of both Tables 15 and 16 permit certain preliminary appraisals. Those recorded in Table 15 indicate that every full program considered shows net advantage in Puerto Rico compared with identical mainland operations. Subsequent differential profitability corrections change the preliminary net-advantage terms to varying extent, primarily because of the scale advantages of mainland complexes.[18] The differential profitability corrections also vary in magnitude from program to program. Nevertheless a consideration of the preliminary net-advantage figures is instructive, particularly when the full programs are contrasted with the short programs.

As already noted, the data of Table 15 indicate that petrochemical activities are at a disadvantage when conducted in Puerto Rico. Thus the short programs selected for investigation were designed so as to include virtually no petrochemical intermediate activities.[19]

As could be expected, most of the short programs achieve a net advantage in Puerto Rico compared with identical mainland operations. More important is the fact that the preliminary net advantage of Short Program A (staple fiber activity alone) in Puerto Rico exceeds by $92,000/yr the greatest preliminary net advantage recorded for any of the full complexes (Dacron A). And for Short Program D, which combines both fiber and fertilizer activities, the preliminary net advantage is $816,000/yr greater than that of

---

[17] The latter two advantage terms are overstated slightly, owing to the fact that the advantages on fiber activities and fertilizer activities reflect fuel-input disadvantages which are less than they should be. (See previous footnote.) The error is approximately $7000/yr.

[18] See p. 155.

[19] An exception in some short programs is the production of ammonia for the fertilizer activities.

PRELIMINARY NET ADVANTAGES

the most advantageous full program. These figures apply under the assumption that the fertilizer and fiber activities of the short programs in Puerto Rico must utilize imported fuel oil and polymer materials.

The data of Tables 15 and 16 indicate that fertilizer activities alone (Short Program B), the refinery activity alone (Short Program C), or a combination of these activities (Short Program E) are definitely less advantageous for Puerto Rico than the best full programs and short programs. However, it is important to note that Short Program F, which combines refinery, fertilizer, and staple fiber activities, compares not too unfavorably with Short Program D, which includes only fiber and fertilizer activities, and which has the highest preliminary advantage term of any program considered. In fact, under the more favorable assumptions regarding transport cost of crude-oil and refinery products, the net advantages of programs F and D are virtually the same. These facts suggest that from the standpoint of Puerto Rico, Short Program F would be definitely preferable to Short Program D under these more favorable assumptions. The addition of the refinery to the fiber–fertilizer complex would not reduce appreciably any component of the latter's advantage, but would achieve for Puerto Rico a larger and more integrated complex of activities with the attendant economic and social benefits of increased investment and employment.

In brief, evaluation of the data of Tables 15 and 16 indicates that, with reference to cost and revenue differentials which can be explicitly computed, the net advantage of a Puerto Rico complex compared with an identical mainland complex is enhanced by synthetic fiber and fertilizer activities. The amount of such net advantage is changed little by the inclusion of refinery activities, but it is decreased by the operation of petrochemical activities. Thus the computed net advantages for Puerto Rico of programs that include fiber and fertilizer activities but omit the petrochemical activities are considerably greater than the computed net advantages of the full programs. This situation is not likely to be changed by the subsequent differential-profitability corrections. A primary component of those corrections is the reduction of Puerto Rico's advantage by amounts representing the greater scale economies achievable on the mainland. And it is in the production of petrochemicals that mainland operations are likely to possess the greatest scale advantages.

Aside from consideration of scale, why are petrochemical activities disadvantageous for Puerto Rico? First, all petrochemical activities require substantial inputs of fuel, not only for direct-process use but also for the generation of steam and power. With respect to fuel inputs, Puerto Rico is at a considerable disadvantage compared to the Gulf Coast. The abundance of natural gas in the latter area holds fuel costs well below the Puerto Rico fuel-oil export price—the price that determines the cost of all fuel on the island.[20] Second, petrochemical activities require considerable inputs of skilled labor. These requirements place Puerto Rico at a substantial disadvantage in petrochemical production, because of the premium wage postulated for such labor in Puerto Rico.

It is true that full complexes in Puerto Rico would avoid the transport costs on petrochemical commodities which would be incurred by complexes importing such products. However, as the data of Tables 15 and 16 indicate, this transport-cost advantage of Puerto Rico petrochemical activities is considerably outweighed by their disadvantages on fuel and labor account.

EFFECTS OF ALTERNATIVE WAGE-RATE DIFFERENTIALS

It is not likely that Puerto Rico's disadvantage on fuel costs will be appreciably reduced, at least in the forseeable future. However, it may reasonably be anticipated that, with increasing industrialization, Puerto Rico's disadvantage on chemical–petroleum labor will be reduced and even eliminated (i.e., Puerto Rico and mainland wages may be equalized for this type labor). In column 2 of Table 17 appear the locational advantages in Puerto Rico of the twenty-eight full programs and the six short programs, calculated on the basis of equalization of Puerto Rico and mainland chemical-petroleum labor costs. As in preceding calculations, the data reflect the comparison of essentially identical Puerto Rico and mainland complexes. (For comparison, the advantages of the programs as originally calculated are recorded in column 1 of Table 17.)[21]

---

[20] Natural gas is relatively low in price at the Gulf Coast because shipment of natural gas by pipeline is more expensive than shipment of a Btu-equivalent amount of fuel oil by tanker.

[21] Except for chemical–petroleum labor inputs, the initial set of price differences underlies the calculations of Table 17. The figures for the short programs which do not include the refinery activity are based on the fuel-oil import price; for the other short programs the figures are based on the fuel-oil export price.

TABLE 17
PRELIMINARY LOCATIONAL ADVANTAGES OF SELECTED FULL AND SHORT PROGRAMS UNDER ALTERNATIVE SETS OF LABOR PRICE DIFFERENCES (IN $M/YR)

Program		Sets of Labor Price Differences		
		CL = +$1.00/mhr [a] TL = -$0.75/mhr [b]	CL = 0 TL = -$0.75/mhr	CL = +$0.50/mhr TL = -$0.375/mhr
		(1)	(2)	(3)
Orlon	A	589	2885	-252
	B	976	3305	147
	C	793	3068	-56
	D	957	3262	119
	E	481	3061	-251
	F	618	3209	-110
	H	643	2822	-243
	J	1093	3310	221
Dynel	A	696	3276	-32
	B	898	3329	108
	C	939	3322	131
	D	895	3526	182
	E	1032	3495	255
	F	1173	3563	368
Dacron	A	1471	3700	604
	B	1235	3708	462
	C	1397	3623	529
	D	1161	3630	386
Nylon	A	974	3598	259
	B	973	3631	271
	C	820	3544	143
	D	708	3227	-48
	E	707	3260	-36
	F	554	3173	-163
	G	927	3202	77
	H	957	3278	125
	J	1087	3752	387
	K	485	3026	-262
Short Program	A	1563	2944	367
	B	724	907	773
	C	-704	-462	-639
	D	2287	3851	1140
	E	67	492	181
	F	1662	3468	580

[a] CL = Price Difference on Chemical-petroleum labor
[b] TL = Price Difference on Textile labor

Although the more favorable assumption on skilled-labor cost definitely improves the locational advantage of Puerto Rico for all the programs considered, more significant is the change in the relative attractiveness of short and full programs. The data of column 2 show that several of the full programs (e.g., Dacron A, Dacron B,

Nylon B, and Nylon J) now achieve virtually as great a locational advantage as the best short program (Program D). Thus, with the disadvantage on chemical labor costs removed, Puerto Rico complexes that include petrochemical activities no longer suffer any appreciable disadvantage compared to short complexes containing only fiber and fertilizer activities. As already indicated, with other things equal, the qualitative social and economic advantages of the full complexes would make them more attractive in Puerto Rico than the short programs.

Puerto Rico's disadvantage on chemical-petroleum labor costs may indeed diminish and eventually disappear. By the same token, her *advantage* on textile labor may be appreciably narrowed in the future. The figures of column 3 of Table 17 are calculated on the basis that the initially postulated price differences on both types of labor are reduced by one-half.[22] As could be expected, for all the full programs and most of the short programs, the net effect is to reduce Puerto Rico's locational advantages when compared to the original calculations. Exceptions are Short Programs C and E, which contain no fiber activities and thus are subject only to the favorable effects of the new set of labor price differences. Short Program D still emerges with the greatest locational advantage of all the programs, but its margin over the best full program, Dacron A, is reduced by over 30%, i.e., from $816,000 to $536,000. Short Program A is most adversely affected, because it involves fiber activity alone.

It is to be emphasized once again that the data on the locational advantages of Puerto Rico complexes summarized in this chapter are preliminary. Calculation in succeeding chapters of differential profitability corrections will change the magnitude of the net advantage terms. Furthermore, corrections in preliminary locational advantage amounts occasioned by differential profitabilities will differ widely from program to program. In fact, such calculations will suggest the desirability of certain programs which have not been selected for analysis in view of the factors investigated thus far. For example, the results of the differential-profitability analysis will point up the advantage for Puerto Rico of a *reduced Dacron A program,* to be discussed below. Nevertheless, in spite of the preliminary character of the findings of this chapter, the

---

[22] That is, Puerto Rico's advantage on textile labor becomes 37.5 cents instead of 75 cents per man-hour; and her disadvantage on chemical-petroleum labor becomes 50 cents instead of $1.00 per man-hour.

PRELIMINARY NET ADVANTAGES 143

consideration of the many alternative short and full programs provides a basis for a more meaningful examination of the differences in the way the various programs are affected by the differential-profitability corrections.

Finally, it is to be noted that a comparison of the preliminary locational advantage of both short and full programs is not strictly valid. As discussed on page 35 of Chapter 2, there may be associated with full programs greater spatial integration economies of the type that are linked to growth of pools of efficient labor and management skills, savings in indirect production costs, and more effective quality control. Also associated with full programs may be larger social gains stemming from desirable influences upon the rate of industrialization, entrepreneurial attitudes, savings and investment habits, etc. Hence, the margins of preliminary locational advantage which short programs have over full programs are more a measure of the minimum magnitude by which the spatial integration and social advantages of the full programs must exceed those of the short programs for full programs to be pursued.

Chapter 7

Differential-Profitability Corrections: General Remarks

Up to this point, the study has been developed within an analytical framework which contains a major abstraction from reality. For each of the selected programs, the advantage or disadvantage of Puerto Rico has been computed by comparing the program in Puerto Rico with an essentially identical program on the United States mainland.[1] The rationale of this procedure has been fully explained in the preceding chapters. In brief, it makes possible the utilization of short-cut computing methods to illustrate the analytical technique as well as to provide results which are in themselves useful for certain purposes.

However, as mentioned several times, a more complete analysis requires the identification and calculation of possible "differential-profitability corrections." It is quite likely that the complex of activities required in any program considered in the previous chapters differs in various ways from the complex of activities by which the commodities of that program would be efficiently produced on the U. S. mainland. Therefore, the differential profitability resulting from such differences must be evaluated in economic terms, and the findings of the previous chapters accordingly corrected.

[1] In some instances, a few nonidentical productive aspects were encompassed because their effects could be expressed in terms of the general framework of revenue and cost differentials.

THE BASIC TYPES OF REGIONAL NON-IDENTITIES: THEIR INTERRELATIONS

Generally speaking, regionally nonidentical aspect of production which are significant to this study can be divided into two broad types: (1) process differences, and (2) product differences. The characteristics of specific differences of either type may vary widely. Many of the chemical intermediates used in synthetic fiber production can be produced from two or more alternate raw materials. They also can be produced by various alternative chemical reactions. On the other hand, two or more processes may use the same raw materials and the same type of reactions but differ in the proportions used of other productive factors such as machinery and labor, steam and power, etc. With respect to products, a mainland complex might produce one or more commodities entirely different from any to be found in a comparable Puerto Rico complex; or the two lists of commodities could be the same but their relative output proportions different.

The condition for differences in processes and products of a mainland productive complex from those of a comparable program for Puerto Rico is that such differences make the mainland complex more efficient. That is, a mainland operation tends to incorporate those differences in processes and products that increase net revenue from the levels that would result from an operation identical with the Puerto Rico program.

Once it is established that a mainland complex would in fact embody certain different processes and products, the question arises as to how to adjust the figure for Puerto Rico's overall advantage, calculated by methods explained in preceding chapters. One method would be to calculate total costs and revenues for each of two mainland operations, one identical with the Puerto Rico program, the other embodying the different and more efficient processes and products. The excess of *net* revenue of the second operation over the first would measure the overstatement of net advantage (or understatement of net disadvantage) of the program for Puerto Rico as originally calculated. The procedure would require knowledge of mainland prices for all basic outputs and of mainland costs for all basic inputs of both complexes.

It is often possible to adopt a less cumbersome procedure. The activities or outputs of the efficient mainland operation which remain (practically speaking) identical with those of the Puerto Rico program do not give rise to any part of the required adjustment in the overall advantage figure. Thus it is necessary to calculate only the costs and revenues (using mainland prices and costs) associated with those processes and products that differ between the Puerto Rico and mainland operations.

In cases of process differences in producing intermediate commodities which are consumed in later productive stages, there is even no need for knowledge or estimation of explicit costs of the intermediates by the alternative methods. The effect of an intermediate process difference can be expressed in terms of differences in requirements for basic inputs of raw materials, utilities, capital services, and labor required by the different processes. Thus the only mainland costs that must be known are those for basic inputs for which requirements differ in the two operations because of process differences.[2] However, the explicit mainland costs of producing a given intermediate by alternative processes often will have been previously estimated as the only way of choosing the process that is most efficient for the mainland operation. If such is the case the analyst can conveniently measure the adjustment of advantage due to a process difference by using the already known difference in mainland production cost of the intermediate by the two relevant processes rather than by calculating the resultant differences in basic inputs and evaluating them in terms of costs.

Although it is conceptually convenient to divide the nonidentical aspects of the Puerto Rico and mainland programs into process and product differences, frequently close interrelations exist between the two. Process differences can cause or be associated with product differences, and vice versa. Fortunately, with respect to most process differences, the interrelations are of such a nature that the effects of the resulting or associated product differences can be included in the evaluation of the process differences themselves. The analyst in effect can ignore the product differences and thus eliminate the computation of the associated revenue differentials. In fact, in the present study the calculation of the differential-profitability corrections abstracts from *all* product differences. The logic and justification of such a procedure will be clarified in the following discussion of each of the specific ways in which a

---

[2] Any excess of the intermediate produced which is not subsequently consumed by other activities of the complex presumably would be marketed to outside users. If the amount of such excess production differs according to the process used, the result is a product difference. To evaluate this difference requires knowledge of the selling price of the product.

DIFFERENTIAL-PROFITABILITY CORRECTIONS 147

given Puerto Rico program might significantly vary from a comparable efficient mainland operation.

DIFFERENCES IN PROCESS REACTION OR RAW MATERIAL FOR A GIVEN COMMODITY OR INTERMEDIATE: ASSOCIATED PRODUCT DIFFERENCES

In the production programs set up as suitable for Puerto Rico, there are three broad groups of activities: (1) the oil refinery activities, (2) the petrochemical activities (including production of the major final fertilizer components), and (3) the fiber-production activities. It is possible that a mainland operation relevant for comparison with any Puerto Rico program would exhibit processes and products different from the latter program in any or all of the three groups.

Oil-Refinery Activities

For each of the Puerto Rico production programs considered in this study the oil refinery setup is taken to be that designated as prototype 4. Because of the number and complexity of a refinery's processes and products, it is clear that the possibilities for an efficient mainland refinery to differ from a prototype 4 refinery are virtually unlimited. A physical process might vary with respect to conditions of temperature or pressure, or in equipment and labor used. A chemical reaction might differ in similar ways and, in addition, with respect to the specific chemicals used to accomplish the reaction. There is always the possibility, too, that a given process in one refinery may have no counterpart at all in the other. Such a situation is most likely to occur if there are product differences between the two refineries, e.g., if one produces a given grade of light fuel oil or distillate not produced at all by the other.

In spite of the fact that considerations of cost and demand could lead to numerous differences of process and of product in a mainland refinery from the activities of refinery prototype 4, no attempt is made in this study to calculate the differential-profitability corrections that would measure the effect of such differences. The refinery setup chosen for an actual Puerto Rico operation could well differ from refinery prototype 4 in much the same ways as would a mainland refinery, especially since it is presumed that the mainland constitutes the principal market for Puerto Rico refinery products. It is likely that under such conditions the net effect of any process and product differences between the Puerto Rico and mainland refineries would be relatively minor. Hence, the net cost and revenue differentials with respect to refinery activities alone would remain of roughly the same magnitude as those calculated in this study for refinery prototype-4 activities.

Petrochemical Activities

The problem of process and product differences in the petrochemical activities requires a more detailed treatment. Some of the advanced intermediates which enter in the production of synthetic fibers can be produced either from petroleum hydrocarbon raw materials (as in all the programs of this study), or from raw materials of a quite different source. For example, nylon salt can be synthesized from cyclohexane and butadiene, which are petrochemicals, or from benzene and furfural, which can be produced from coal and coke by-products and vegetable waste materials, respectively. In many cases, Puerto Rico lacks economic access to such alternative raw materials. If it is found more efficient on the mainland to produce a given chemical intermediate from raw materials other than petroleum hydrocarbons, the result is a significant process difference from a Puerto Rico operation limited to the petrochemical route.

Moreover, in the production of many intermediates, alternative raw materials may be used, all of which can be derived basically from petroleum. For example, among other ways nylon salt can be produced entirely from cyclohexane, entirely from butadiene, or from a combination of the two. Nevertheless, the production processes are often quite different in the type and number of reactions required, severity of reaction conditions, amounts of process chemicals used, etc. Such differences, too, must be evaluated if they in fact would exist between a given Puerto Rico program and an efficient mainland operation.

In order to be essentially relevant, a differential-profitability correction for a process difference must be based on a comparison of (1) the process assumed for the specific Puerto Rico program being considered, with (2) the process that would be employed in a similar mainland operation. It is assumed in this study that the major criterion for choice among alternative mainland processes tends to be cost. Accordingly, for each chemical intermediate re-

DIFFERENTIAL-PROFITABILITY CORRECTIONS 149

quired in the production of the synthetic fibers and fertilizer components listed in the possible Puerto Rico programs, Gulf Coast costs of production have been estimated for all economically feasible production processes. The lowest-cost process is then utilized in all subsequent calculations of differential profitability associated with that particular chemical intermediate. If a given Puerto Rico program includes production of the intermediate by a method other than the lowest-cost mainland process, the differential-profitability correction can be computed by multiplying the difference in mainland unit production costs of the two processes by the amount of the chemical intermediate required by the program.

Often a chemical intermediate of the type used directly in synthetic fiber activities is produced by successive reaction or combination of more basic intermediates. In many cases there are alternative processes and materials possible for achieving each of the successive intermediates. Therefore, the lowest mainland production cost of a synthetic fiber intermediate can be found only after the calculation of the lowest-cost processes for making each of the more basic intermediates. For this reason the calculation in this study of differential-profitability corrections due to process differences has utilized directly the cost differentials of the intermediates rather than trace back the resultant differences in basic raw materials, utilities, etc.

Important process differences of the type just discussed, i.e., use of different raw materials or of different reactions which involve different raw material or utility proportions, almost always are accompanied by certain kinds of *product* differences. For example, a Puerto Rico program for producing Orlon fiber might utilize ethylene oxide to make the required acrylonitrile, while a mainland producer might find the acetylene route to acrylonitrile cheaper, provided the acetylene were manufactured from calcium carbide rather than petroleum hydrocarbons. The mainland operation would use none of its hydrocarbon streams to produce ethylene (or acetylene) and thus would have greater amounts of such streams available as *products* than would the Puerto Rico operation.

Fortunately, the effects of product differences which are direct consequences of a process difference need not be evaluated specifically. They are included in the evaluation of the process difference itself. In the above example the "saving" of the mainland producer in using the cheaper acetylene process includes the extra revenue he can obtain from selling or otherwise using the additional hydrocarbon streams.

There are other possible types of product differences in the petrochemical activities. Compared with a Puerto Rico program producing Dacron and thus using ethylene glycol, a mainland producer might find it more profitable to utilize the required ethylene in producing polyethylene, a plastic material. However, in this study no attempt is made to discover or evaluate such product differences. It is assumed that, generally speaking, petrochemical raw materials will be ample enough to supply, at least partially, all their economically important uses, and that competition will tend to equalize the profitability of those alternatives.

One other type of product difference requires mention. Conceivably, a mainland operation could increase its total net revenue by expanding the number and diversity of its productive activities. An oil refinery could add petrochemicals, plastics, and fibers; then branch out into synthetic rubber, automotive chemicals, and detergents; and finally move into such diverse fields as metal fabricating and marine transport; etc. It is quite possible that each separate field would be profitable, thus adding to the net revenue of the entire operation. Obviously, because of conditions of demand, materials, and location, a Puerto Rico operation would be precluded from many of the activities available to a mainland operation. Each successive activity added on the mainland complex would in effect create additional process and product differences. The additional net revenue to the mainland operation would eventually overcome any initial advantage possessed by a Puerto Rico program.

Such a situation does not necessarily require a true differential-profitability correction, however. *Only that part of the additional net revenue stemming from economies of integration not available to the extra activities by themselves could be a valid part of such a correction.* Such economies are extremely difficult to identify and to evaluate quantitatively, and lead to subjective welfare considerations. In any case, this class of product and process differences is judged outside the scope of this study.

Fiber Activities

In the final production stages of specific synthetic fibers such as nylon, Orlon, Dynel, and Dacron, the possibilities for important process differences are quite limited. There can be no difference in final intermediates; e.g., the intermediate for Orlon is virtually by definition acrylonitrile. Although detailed descriptions of the

DIFFERENTIAL-PROFITABILITY CORRECTIONS 151

actual processes are not generally available, it appears doubtful that there could be major differences in reactions or in the technology of the fiber-production stage of any specific fiber. As in the other groups of activities, possible minor differences in proportions of different utilities used or of other productive factors are ignored, since the resulting differences in basic inputs would tend on balance to be relatively insignificant.

Since for any given program there is only one fiber product specified, the only product difference possible (aside from a difference in scale of output, to be discussed in following sections) would be a diversification by a mainland operation of its type of fiber output. The reasons for abstracting from the possibility of such product diversification are the same as those cited above in the discussion of possible diversification of petrochemical products.

SCALE DIFFERENCES

Up to this point the discussion has not dealt with a possible regional nonidentity which can result in major differential profitability corrections—the scale of operation. Clearly it is possible that an efficient mainland operation might differ from any given Puerto Rico program in the scale of virtually any of its activities. Actually, scale differences can be described or defined in terms of process and product differences. However, the potential importance of scale differences is so great that they are discussed separately from other process and product differences.

Refinery Activities

As explained previously, virtual identity in process reactions and in product-mix proportions has been assumed in this study to apply to the refinery activities of a Puerto Rico program and the corresponding mainland complex. This postulate in itself does not preclude differences in the overall scale of the refinery and hence in the scale of each refinery activity. However, the scale of the refinery as it appears in the various Puerto Rico programs may well characterize the scale of at least some U. S. mainland refineries for at least the foreseeable future.[3] For this reason, it was considered justifiable to abstract from the possibility of refinery scale differences and their corresponding differential profitabilities.

[3] See also the previous discussion of this point in Chapter 4, p. 63.

Petrochemical Activities

Economies of large-scale operation are pronounced in most petrochemical processes. Within significant ranges, initial capital cost does not rise proportionally with increases in capacity. Since plant and equipment investment cost is high, and consequently fixed charges are large compared to other elements of production cost, it is clear that important economies of scale can be achieved. Furthermore, most petrochemical processes are of the type that require a decidedly less-than-proportional increase in direct labor requirements for any given increase in capacity. This tends to increase economies of scale.

Essentially differences in scale are process differences. Within a specified range of operations, per unit output, a larger-scale activity requires a smaller ratio of (a) direct labor and capital services to (b) direct inputs of raw materials, process materials, and utilities than the same activity on a smaller scale. However, there are also product differences associated with differences of scale in the petrochemical activities. If a petrochemical intermediate is produced on a scale large enough to achieve most of the possible scale economies, the quantity produced will, generally speaking, be considerably greater than that required by a subsequent fiber activity of a size that can be reasonably expected to exist. The result is an excess of the chemical intermediate, which presumably must be sold on the market or by contract. Such a situation could easily exist in a mainland operation. However, it could not arise in a Puerto Rico program since it has been assumed that chemical intermediates would be produced in Puerto Rico only in amounts required internally by a program. Thus the mainland operation would have some quantity of the chemical intermediate as part of its final product mix, whereas the Puerto Rico program would not.

A complete differential-profitability correction would involve computation of the additional revenue from sale of the excess chemical intermediate, and the additional total costs involved in its production. The additional *net* revenue would represent the value of the differential-profitability correction. However, through the use of a simplifying assumption, which actually may do little violence to reality, the adjustment necessary because of scale differences in chemical intermediate production can be estimated more directly. It can be approximately stated in terms of the effect of scale economies in lowering the unit cost of the final chemical intermediates used in producing synthetic fibers and fertilizer. The necessary assumption is that the mainland complex can dispose of its excess production of chemical intermediates only at an approximate break-even price. Then, the correction due to mainland economies of scale in the production of chemical intermediates for fiber production can be calculated by multiplying the quantity of each final intermediate used in the fiber-production activity by the difference between that intermediate's unit production cost computed on the basis of (a) activity scales as given in the original program and (b) activity scales anticipated for the mainland complex. The correction for economies of scale in fertilizer components can be computed in a similar manner. The differences in production cost of final intermediates and components are of course influenced by the economies of scale achieved in the manufacture of all the more basic intermediates used in their production.

Fiber Activities

Empirical data on current and proposed plant capacities in the synthetic fiber industry, together with a consideration of market conditions and other economic factors, point to the conclusion that the industry on the mainland will continue to be characterized by many units in roughly the same size range as those depicted in the Puerto Rico programs. Therefore, in this study there is no calculation of a differential-profitability correction due to mainland economies of scale in the fiber activities.

TECHNIQUE FOR CALCULATING TOTAL DIFFERENTIAL-PROFITABILITY CORRECTIONS

The preceding discussion has indicated that the need for differential-profitability corrections may arise because of a variable and potentially very large number of possible *process* and *product* differences between any given Puerto Rico program and a similar mainland operation. However, it has also been demonstrated that a large number of such differences can justifiably be ignored. In fact, in this study, *all* probable differences in the refinery activities and the synthetic fiber activities are assumed to have minor net effects and are, therefore, ignored.

Even with respect to the remaining group of activities, those for petrochemical production, *product* differences either may be ignored or their specific evaluation avoided. Thus the total differential-profitability correction for each specific program is achieved by an evaluation of *process* differences in the petrochemical activities. These differences may be in the process reaction or in raw materials used, or they may result from differences in activity scales.

In the actual computation, a combined evaluation of the two types of process differences is made. Essentially the procedure is to compute for each selected program the unit costs of production of the final chemical intermediates and fertilizer components for a mainland complex identical with the assumed Puerto Rico complex. Then for the same activities unit costs of production are estimated for a mainland complex, using the most efficient productive processes at scales assumed possible to achieve. The various differences in unit production costs multiplied by the final quantity requirements of the program yield the total amount by which the net advantage of Puerto Rico as originally calculated must be decreased to take into account the more efficient and larger-scale mainland operations.

Computations are made on the basis of three separate assumptions with respect to the probable scale disadvantages of Puerto Rico in the production of petrochemical intermediates. The assumptions are:

1. That mainland petrochemical activities will continue to be of roughly the same scales as are currently common, resulting in "moderate" scale disadvantages for Puerto Rico.

2. That mainland petrochemical activities will typically achieve scales considered optimum by engineering authorities, resulting in "maximum" scale disadvantages for Puerto Rico.

3. That mainland conditions will warrant the existence of self-contained complexes with virtually the same scales for most activities as in Puerto Rico, resulting in "minimum" scale disadvantages for Puerto Rico.

In the chapter immediately following, the differential-profitability analysis for the nylon programs is presented in detail. Since the corrections for the Orlon, Dynel, and Dacron programs are calculated in the same manner as those for nylon, the succeeding chapter, Chapter 9, merely summarizes the results in the Orlon, Dynel, and Dacron programs. In Chapter 10, the final chapter, the overall findings and implications of the differential-profitability analysis are discussed, together with possible modifications and extensions of the analysis.

Chapter 8

Differential-Profitability Analysis: Nylon

It was stated in the preceding chapter that, in the calculation of differential-profitability corrections, differences in the oil-refining and fiber-production operations may be ignored. Likewise, differences in product composition or proportions may be ignored or treated indirectly as a part of the effect of process differences. Thus the problem basically is to evaluate the effects of differences in the production processes or scales which may characterize the petrochemical intermediate activities in the selected Puerto Rico programs and in their mainland counterparts.

It has been previously pointed out that the differential-profitability correction applicable to a specific Puerto Rico program can be calculated by comparing (1) the production costs of the required final fiber and fertilizer intermediates in a mainland operation identical with the Puerto Rico program, with (2) their production costs in a mainland program utilizing the most efficient processes and operating at scales assumed feasible to achieve. Necessarily involved are comparative production-cost estimates for the more basic intermediates that may be required to produce the final intermediates. This analytical procedure is applied first to the nylon programs.

NYLON SALT PRODUCTION COSTS: MAINLAND OPERATIONS IDENTICAL WITH PUERTO RICO PROGRAMS

A logical way to begin the analysis is to present the results of a detailed estimate of *nylon salt* production costs in mainland operations identical with the selected Puerto Rico nylon programs.[1] Table 18 summarizes these estimates. The nylon salt plant capacity recorded in the first row of Table 18 is the same, 36.5 MM lb/yr, for every program because the fiber output set for each program is the same.

Plant-Investment Costs

The plant-investment figures in the second row of Table 18 require comment. Although there are many different combinations of raw materials and productive processes which can lead to nylon salt, published data and other available information concerning plant-investment requirements are of a very general and limited nature. For that reason, the same figure is assigned to all processes using cyclohexane, furfural, or butadiene as raw materials. The processes utilizing benzene are assigned a slightly higher investment cost because the benzene must first be converted to cyclohexane, after which the process is similar to the others. The lower figure, $28 MM, is derived from estimates by Airov.[2]

Raw-Material Cost Rates

The several rows of Table 18 that refer to raw materials show that several of them are costed at one rate per pound for some programs and at another rate per pound for others. Such differences result from the varying requirements for these raw materials in different programs, which in turn lead to scale differences in the production of the raw materials. Thus the application of correct rates requires knowledge of the output scales for each raw-material commodity produced for use within any given program. The first four columns of Table 19 list by programs the required output scales of various raw-material commodities. Such materials as

---

[1] Nylon salt is the final chemical intermediate for nylon fiber activities.

[2] Joseph Airov, *The Location of the Synthetic Fiber Industry*, John Wiley & Sons, 1959.

TABLE 18
PRODUCTION COST OF NYLON SALT, $/100 LB, PROGRAMS IDENTICAL WITH PUERTO RICO

Programs:		A	B	C	D	E	F	G	H	J	K
Plant Size: (MM lb/yr)		36.5	36.5	36.5	36.5	36.5	36.5	36.5	36.5	36.5	36.5
Plant Investment: (MM $)		28	28	29.5	28	28	29.5	28	28	28	28
Raw Materials:											
85% Cyclohexane,	@ 2.628¢/lb	2.865	2.129		2.865	2.129					
Furfural,	@ 5.256¢/lb	4.100	4.100	4.100				7.122	5.282	8.304	5.121
Butadiene,	@ 6.566¢/lb			1.508	2.561	2.561	2.561				
Benzene,	@ -2.320¢/lb						1.508				
HCl (100%),	@ 4.748¢/lb									4.463	
Sodium Cyanide,	@ 6.323¢/lb	2.972	2.972	2.972						6.097	
Hydrogen,	@ 5.978¢/lb	3.437	3.437	3.437							
Ammonia,	@ 6.739¢/lb	0.036	0.036	0.081	0.045	0.045	0.090	0.036	0.036	0.036	0.054
Sulfuric Acid,	@ 1.933¢/lb							0.348	0.348		
Hydrogen Cyanide,	@ 0.850¢/lb				1.404	1.404	1.404			0.408	0.408
Limestone,	@ 4.842¢/lb										2.492
Chlorine,	@ 0.200¢/lb				0.106	0.106	0.106				0.214
	@ 4.296¢/lb				1.711	1.711	1.711				
Nitric Acid,	@ 3.566¢/lb										
	@ 2.600¢/lb		1.676	1.676		1.676	1.676		3.901		2.496
Utilities:											
Steam,	@ 40.0¢/Mlb	1.262	1.226	1.566	1.252	1.216	1.556	1.544	1.454	1.286	1.264
Cooling Water,	@ 1.5¢/M gal	0.126	0.133	0.218	0.132	0.150	0.227	0.227	0.253	0.135	0.225
Electric Power,	@ 0.5¢/kwh	0.200	0.235	0.280	0.200	0.235	0.280	0.305	0.395	0.180	0.175
Fuel,	@ 15.0¢/MMBtu	0.087	0.066	0.066	0.070	0.048	0.048	0.120	0.068	0.085	0.049
Direct Labor:	@ $2.50/mhr	1.662	1.830	1.975	1.802	1.970	2.115	1.270	1.435	1.587	1.753
Capital and Indirect Costs:											
Supervision		0.166	0.183	0.198	0.180	0.197	0.212	0.127	0.144	0.159	0.175
Plant Maintenance		3.068	3.068	3.233	3.068	3.068	3.233	3.068	3.068	3.068	3.068
Equipment and Operating Supplies		0.460	0.460	0.485	0.460	0.460	0.485	0.460	0.460	0.460	0.460
Payroll Overhead		0.504	0.532	0.568	0.527	0.555	0.591	0.439	0.467	0.492	0.569
Indirect Production Cost		2.142	2.216	2.356	2.204	2.278	2.418	1.970	2.043	2.110	2.182
Depreciation		7.670	7.670	8.082	7.670	7.670	8.082	7.670	7.670	7.670	7.670
Taxes		0.767	0.767	0.808	0.767	0.767	0.808	0.767	0.767	0.767	0.767
Insurance		0.767	0.767	0.808	0.767	0.767	0.808	0.767	0.767	0.767	0.767
Interest		3.068	3.068	3.233	3.068	3.068	3.233	3.068	3.068	3.068	3.068
Total		35.359	36.571	37.650	30.859	32.081	33.152	29.308	31.626	41.142	32.977
Less By-product Credit						.161	.161			1.300	1.655
Final Total		35.359	36.571	37.650	30.698	31.920	32.991	29.308	31.626	39.842	31.322

TABLE 19

ACTIVITY SCALES FOR PUERTO RICO NYLON PROGRAMS AND FOR COMPARABLE MAINLAND PROGRAMS, BY SCALE ECONOMIES ASSUMPTION (IN MM LB/YR)

Activity	Scales for Puerto Rico				Scales for Minimum Scale Economies					Scales for Moderate Scale Economies		Scales for Maximum Scale Economies
	Programs A,D,G	Programs B,C,E,F	Programs H,L	Programs J,K	Programs A,D,G M,N,D	Programs B,C,E,F	Programs H,L	Programs J,K		Programs H,L	All other Programs	All Programs
Nylon Salt	36.5	36.5	36.5	36.5	36.5	36.5	36.5	36.5		36.5	36.5	120
Hydrogen Chloride	20	20		35	100	100		100			100	100
Sodium Cyanide	20	20		40	20	20		40				100
Hydrogen Cyanide	10	10		20	10	10		20			100	100
Chlorine	20	20		35	66	66		66			66	66
Sulfuric Acid				20	100	100		100			100	100
Ammonia	100	100	100	100	100	100	100	100	140		140	140
Nitric Acid	70	100	140	70	70	100	140	70	140		100	200
Ammonium Nitrate	90	90	90	90	90	90	90	90	100		100	200
Urea	70	70	70	70	70	70	70	70	100		100	200

DIFFERENTIAL-PROFITABILITY: NYLON

furfural, cyclohexane, butadiene, and benzene are not listed in Table 19 because their production is assumed not to take place within the programs being analyzed. Their output scales are thus essentially independent of the demands of any given nylon program.

The necessity for specific per-pound rates on the raw-material commodities requires for most of them a calculation of their production costs. Appendix C contains tables of estimated production cost for these various basic and intermediate commodities at various output scales.

Use of Tables in Appendix C to determine Raw-Material Rates

An example will clarify the use of the tables in Appendix C. Suppose one wishes to know the rate at which to cost the sodium cyanide requirement for the program Nylon A. Sodium cyanide requires as a principal intermediate hydrogen cyanide, which in turn requires ammonia in its production. Table 19 shows that the output scales of these three commodities required by Nylon A are, in millions of pounds per year, sodium cyanide, 20; hydrogen cyanide, 10; and ammonia, 100.

Nylon A calls for the total ammonia requirements to be produced partly from the hydrogen-rich re-former gas, partly from the refinery-gas hydrocarbons, and partly from fuel oil. Table C-1 in Appendix C for ammonia-production costs reveals the total estimated cost, in dollars per hundred pounds (or cents per pound), of ammonia from each of these raw materials at output scales of 100 MM lb/yr (as well as at other scales). Since the ammonia-production process can utilize a mixture of raw materials, it is valid to estimate the weighted average cost by using the proportions of the raw materials designated in the program. This weighted average cost for ammonia proves to be 1.933¢/lb.

The next step is to note the cost of the hydrogen cyanide intermediate (Table C-14, Appendix C). At a scale of 10 MM lb/yr, and with ammonia at 1.933¢/lb, hydrogen cyanide is seen to cost 4.842¢/lb. Finally, Table C-18 of Appendix C indicates a sodium cyanide cost of 6.739¢/lb, from a plant producing 20 MM lb/yr and using hydrogen cyanide at 4.842¢/lb. The figure of 6.739¢/lb can then be used to compute the cost, $3.437, of the sodium cyanide input required in the production of 100 lb of nylon salt in the program Nylon A.

In a similar fashion, it can be estimated that for the program Nylon J, which requires larger output scales for both sodium cyanide and hydrogen cyanide, the rate at which sodium cyanide should be costed is 5.978¢/lb. It will be noted that for a few commodities, e.g. limestone and hydrogen in Table 18, caustic soda in Table C-18 of Appendix C, and natural gas (methane) in Table C-14 of Appendix C, there are no tables of estimated production costs. Limestone is a low-cost, large-volume commodity which is not likely to be produced within a complex of the type dealt with in this study. It is therefore costed at its listed selling price as of early 1956. Hydrogen, caustic soda, and methane may all be produced within the complex being analyzed. However, they are essentially by-products, and as such their cost reflects their value in alternative uses, rather than as cost of production. Thus hydrogen and methane are costed at their fuel value (at a rate of 15¢/MM Btu), and caustic soda at its 1956 listed selling price.

It might be argued that, since the basic nylon raw materials—cyclohexane, furfural, butadiene, and benzene—are assumed in all the nylon programs to come from outside producers, they should simply be costed at their listed selling prices. However, for such commodities the listed selling prices often bear little relation to actual production costs, although contract purchasers may be quoted prices that conform much more closely to costs. Therefore, production costs are estimated for benzene and butadiene from large-scale plants of a size either already in existence or reasonably likely in the future. In all nylon programs, these estimates are the basis for costing benzene and butadiene.

Because of the lack of available material and investment-cost information, it is impossible to prepare detailed estimates of the production cost of furfural and cyclohexane. However, rough estimates are made by calculating the ratio between the estimated production cost of butadiene and its listed selling price, and then applying that ratio to the listed selling prices of furfural and cyclohexane. This procedure may not be as arbitrary as it would appear. A check of the ratios of estimated production cost to listed selling price for several other commodities, such as benzene and ammonia, reveals a surprising degree of similarity to the butadiene ratio.

Utility, Direct Labor, and Capital and Indirect Costs

The utility and wage rates used, as well as the method of estimating capital and indirect costs, are discussed in Appendix C. These rates apply generally to all production-cost estimates used in calculating differential-profitability corrections.

NYLON SALT PRODUCTION COSTS: MAINLAND OPERATIONS ACHIEVING MAXIMUM SCALE ECONOMIES

When the investigator has established estimated production costs of nylon salt in mainland operations identical with the selected Puerto Rico programs, his next task is to estimate nylon salt production costs in a mainland operation utilizing the most efficient productive processes run at the most efficient possible scales. Table 20 summarizes the results of such estimates under the assumption that the mainland operation can achieve "maximum" scale economies. As in the case of Table 18, the rate at which the various raw-materials inputs should be costed can be determined from examination of the tables in Appendix C and of the final column of Table 19, which indicates the output scales for the activities within the complex.

Choice of Output Scales for Specific Activities

The selection of 120 MM lb/yr as the "maximum" scale for nylon salt is based on empirical and analytical considerations cited by Airov,[3] and on other general technological, locational, and market considerations. The scales selected for the other activities follow the empirical findings and analytical conclusions of Isard and Schooler,[4] modified where necessary by later developments and trends.

It should be pointed out that the output scales for the raw-material commodities are not necessarily set by the requirements of the program as they are in most cases in the Puerto Rico programs. The greater and more varied demands for industrial chemicals on the mainland are assumed to allow large-scale production of each, regardless of the magnitude of the demand by nylon producers.

For commodities whose cost is determined by their by-product value (e.g., refinery-gas hydrogen and methane), and commodities such as cyclohexane and butadiene, which are produced outside the complex, the rates are the same as those used in the preparation of Table 18. Obviously their cost is for practical purposes independent of the activity scales within the nylon-production complex.

[3] *Ibid.*
[4] *Location Factors in the Petrochemical Industry*, U. S. Department of Commerce, Office of Technical Services, 1955.

INDUSTRIAL COMPLEX ANALYSIS

TABLE 20
PRODUCTION COST OF NYLON SALT

Programs:		A	B	C	D	E
Plant Size: (MM lb/yr)		120	120	120	120	120
Plant Investment: (MM $)		57	57	60	57	57
Raw Materials:						
85% Cyclohexane,	@ 2.628¢/lb	2.865	2.129		2.865	2.129
Furfural,	@ 5.256¢/lb	4.100	4.100	4.100		
Butadiene,	@ 6.566¢/lb				2.561	2.561
Benzene,	@ 2.320¢/lb			1.508		
HCl (100%),	@ 3.223¢/lb	1.515	1.515	1.515		
Sodium Cyanide,	@ 5.150¢/lb	2.626	2.626	2.626		
Hydrogen,	@ 0.9¢/lb	0.036	0.036	0.081	0.045	0.045
Ammonia,	@ 1.880¢/lb					
Sulfuric Acid,	@ 0.683¢/lb					
Hydrogen Cyanide,	@ 3.430¢/lb				0.995	0.995
Limestone,	@ 0.200¢/lb				0.106	0.106
Chlorine,	@ 1.589¢/lb				0.763	0.763
Nitric Acid,	@ 1.850¢/lb		1.480	1.480		1.480
Utilities:						
Steam,	@ 40.0¢/M lb	1.262	1.226	1.566	1.252	1.216
Electric Power,	@ 0.5¢/kwh	0.200	0.235	0.280	0.200	0.235
Cooling Water,	@ 1.5¢/M gal	0.126	0.133	0.218	0.132	0.150
Fuel,	@ 15.0¢/MMBtu	0.087	0.066	0.066	0.070	0.048
Direct Labor:	@ $2.50/mhr	0.658	0.710	0.765	0.713	0.765
Capital and Indirect Costs:						
Supervision		0.065	0.071	0.077	0.071	0.077
Plant Maintenance		1.900	1.900	2.000	1.900	1.900
Equipment and Operating Supplies		0.285	0.285	0.300	0.285	0.285
Indirect Production Cost		1.163	1.186	1.257	1.188	1.211
Payroll Overhead		0.251	0.260	0.277	0.260	0.269
Depreciation		4.750	4.750	5.000	4.750	4.750
Taxes		0.475	0.475	0.500	0.475	0.475
Insurance		0.475	0.475	0.500	0.475	0.475
Interest		1.900	1.900	2.000	1.900	1.900
Total		24.739	25.558	26.116	21.006	21.835
Less By-product credit					0.161	0.161
Final Total		24.739	25.558	26.116	20.845	21.674

Estimated Production Cost, by Process

Table 20 contains estimates of the production cost of nylon salt by fourteen different processes. The first ten are the same as those in the programs set up for Puerto Rico, whereas the last four are additional possible programs based wholly or partially on benzene as a raw material.[5] It is felt that these fourteen processes

---

[5] Details of the productive processes employed in nylon programs A through K appear in Appendix B.

Program L calls for the production of nylon salt via the all-adipic-acid route, with the adipic acid produced from benzene via nitric acid oxidation of cyclohexane. Pro-

DIFFERENTIAL-PROFITABILITY: NYLON

TABLE 20
$/100 LB: MAXIMUM SCALE ECONOMIES

F	G	H	J	K	L	M	N	O
120	120	120	120	120	120	120	120	120
60	57	57	57	57	60	60	60	60
	7.122	5.282						
			8.304			4.100		
2.561				5.121			2.561	
1.508					3.712	2.018	2.018	5.011
			3.030		1.515			
			5.253			2.627		
0.090	0.036	0.036	0.036	0.054	0.144	0.099	0.108	0.189
	0.338	0.338			0.338			0.338
			0.328	0.328				
0.995				1.989			0.995	
0.106				0.214			0.106	
0.763				0.525			0.763	
1.480		3.663			3.663			
1.556	1.544	1.454	1.286	1.264	2.299	1.722	1.712	2.683
0.280	0.305	0.395	0.180	0.175	0.510	0.260	0.260	0.465
0.227	0.227	0.253	0.135	0.225	0.381	0.217	0.217	0.385
0.048	0.120	0.068	0.085	0.049	0.068	0.087	0.070	0.120
0.823	0.503	0.568	0.658	0.693	0.635	0.718	0.773	0.575
0.082	0.050	0.057	0.066	0.069	0.064	0.072	0.077	0.058
2.000	1.900	1.900	1.900	1.900	2.000	2.000	2.000	2.000
0.300	0.285	0.285	0.285	0.285	0.300	0.300	0.300	0.300
1.282	1.095	1.124	1.164	1.179	1.200	1.236	1.260	1.173
0.286	0.225	0.236	0.251	0.257	0.255	0.269	0.278	0.245
5.000	4.750	4.750	4.750	4.750	5.000	5.000	5.000	5.000
0.500	0.475	0.475	0.475	0.475	0.500	0.500	0.500	0.500
0.500	0.475	0.475	0.475	0.475	0.500	0.500	0.500	0.500
2.000	1.900	1.900	1.900	1.900	2.000	2.000	2.000	2.000
22.387	21.350	23.259	30.561	22.927	23.569	25.240	21.498	21.542
0.161			1.300	1.655			.161	
22.226	21.350	23.259	29.261	21.272	23.569	25.240	21.337	21.542

include virtually all the raw material and process combinations that could be commercially feasible in producing nylon salt. Examination of the final total row of Table 20 shows that Nylon D is the lowest-cost program, with a cost per hundred pounds of nylon salt of $20.845.

---

gram M requires the adipic acid—adiponitrile route to nylon salt, with adipic acid from benzene via air oxidation of cyclohexane, and adiponitrile from furfural. Program N is the same as Program M, except the adiponitrile is produced from butadiene. Program O employs the all-adipic route to nylon salt, with the adipic acid produced from benzene via air oxidation of cyclohexane.

DIFFERENTIAL-PROFITABILITY CORRECTIONS: MAXIMUM MAINLAND ECONOMIES OF SCALE

Correction for Nylon Salt

The next step is to compute the differential-profitability correction for the Puerto Rico programs due to nylon salt production. It is accomplished by subtracting 20.845[6] from the final total figure for each program as shown in Table 18 and multiplying the result (representing cents per pound) by 36.5 MM, the yearly output in pounds of nylon salt. For example, for the program Nylon A, $35.359 - 20.845 = 14.514$. This figure, multiplied by 36.5 MM is equal to $5,297,610. This is the amount by which the preliminary locational-advantage figure for program Nylon A, obtained in the first part of this study, must be reduced to take account of more efficient and larger-scale mainland operations in the production of nylon salt.

Corrections for Fertilizer Commodities

But there are other corrections also to be made. In addition to nylon salt, the petrochemical activities section of each program produces the fertilizer commodities: ammonium nitrate and urea. Urea requires ammonia as an intermediate, and ammonium nitrate is made from ammonia and from nitric acid, which in turn is based on ammonia. For Nylon A, the first column of Table 19 shows the required output scales, in millions of pounds per year, to be ammonia, 100; nitric acid, 70; ammonium nitrate, 90; and urea, 70 (when mainland scales are identical with Puerto Rico scales). The corresponding scales in the optimum mainland operation are, respectively, 140, 200, 200, and 200, as given in the last column of Table 19. It has already been shown that the rate at which ammonia should be costed in a mainland operation identical with Nylon A is $1.933/100 lb. Starting with this figure and utilizing the Appendix C cost tables for nitric acid, ammonium nitrate, and urea, we find that ammonium nitrate and urea can be produced in the mainland operation identical with Puerto Rico program Nylon A for $2.590 and $2.473/100 lb, respectively. Similarly, the rates for the other nylon programs can be determined.

---

[6] The amount of 20.845 is subtracted since this is the cost in cents of producing one pound of nylon salt on the mainland by the cheapest method.

The rate at which ammonia is costed in the optimum-scale mainland operation requires explanation. According to Table C-1, Appendix C, the lowest-cost process at the scale of 135 MM lb/yr (as well as at smaller scales) is the one based on hydrogen-rich reformer gas. However, the supply of such gas is completely inadequate to sustain the volume production of ammonia in the United States. Therefore, the next lowest-cost process, based on natural-gas or ordinary refinery-gas hydrocarbons which are in adequate supply, tends to determine the relevant mainland rate for ammonia. Thus, ammonia is costed in the optimum-scale mainland operation at a rate of $1.88/100 lb. The rates for ammonium nitrate and urea, per hundred pounds, are then found to be $2.107 and $2.098, as calculated, respectively, in Tables C-3 and C-4.

The differential-profitability correction for Nylon A due to ammonium nitrate production is computed by subtracting 2.107 from 2.590 and multiplying the result (considered as cents per pound) by 87 million, which is the actual output of ammonium nitrate required by Nylon A. The amount of the correction proves to be $420,210. By a similar procedure, the correction for Nylon A due to urea production is calculated to be $258,750.

Total Combined Correction

Addition of the corrections due to nylon salt, ammonium nitrate, and urea yields a total differential-profitability correction for Nylon A of $5,976,570/yr. The locational advantage figure for Nylon A, computed previously, is only $974,000/yr. Therefore the program has a net overall *dis*advantage in Puerto Rico of $5,002,570/yr compared to a mainland operation utilizing the most efficient processes run at their optimum output scales.

Evaluation of Results

By exactly similar procedures, the total correction for each nylon program is found, and in turn the net overall advantage or disadvantage of each program is calculated. These results are summarized in the upper section of Table 21. Inspection of the figures shows that all ten nylon programs suffer very considerable net overall disadvantages. By far the most important reason is the disadvantage due to nylon salt production. The significantly larger scale assumed for mainland nylon salt plants, together with the very high investment and capital costs for nylon salt production

TABLE 21

SUMMARY OF ADVANTAGES AND DISADVANTAGES
(IN $/YR)

		Nylon A		Nylon B	
Mainland Scale Economies	**Maximum**				
	Preliminary Net Locational Advantage:		974,000		973,000
	Process and/or Scale Disadvantages:				
	Nylon Salt	5,297,610		5,739,990	
	Ammonium Nitrate	420,210		264,480	
	Urea	258,750	5,977,000	258,750	6,263,000
	Net Overall Advantage: Disadvantage		5,003,000		5,290,000
	Moderate				
	Preliminary Net Locational Advantage:		974,000		973,000
	Process and/or Scale Disadvantages:				
	Nylon Salt	2,212,265		2,654,645	
	Ammonium Nitrate	192,270		36,540	
	Urea	138,690	2,543,000	138,690	2,830,000
	Net Overall Advantage: Disadvantage:		1,569,000		1,857,000
	Minimum				
	Preliminary Net Locational Advantage		974,000		973,000
	Scale Advantage:				
	Ammonium Nitrate			155,730	1,129,000
	Process and/or Scale Disadvantages:				
	Nylon Salt		2,209,000		2,651,000
	Net Overall Advantage: Disadvantage:		1,235,000		1,522,000

compared with other chemicals, results in large absolute economies for the mainland plants. Of course, many of the Puerto Rico programs suffer additional disadvantage because they are based on the comparatively higher-cost nylon salt processes. As Tables 18 and 20 show, there are considerable production-cost differences among the various processes, at *any* scale. However, it is interesting to note that the program Nylon D, which utilizes the process of lowest cost under conditions of large mainland scales, has only the third lowest overall disadvantage among the Puerto Rico programs —a disadvantage that exceeds those of both Nylon G and Nylon H. This situation results partly because, as output scales diminish, Nylon D's unit nylon salt costs rise faster than those of G and H; partly because H suffers fewer scale diseconomies in fertilizer component production than D; and partly because D's preliminary locational advantage figure is less than those of G and H.

TABLE 21
INDIVIDUAL PUERTO RICO NYLON PROGRAMS, BY MAINLAND SCALE ECONOMIES ASSUMPTION

Nylon C		Nylon D		Nylon E	
820,000		708,000		707,000	
6,133,825		3,596,345		4,042,375	
264,480		420,210		264,480	
258,750	6,657,000	258,750	4,275,000	258,750	4,566,000
	5,837,000		3,567,000		3,859,000
	820,000		708,000		707,000
3,048,480		511,000		957,030	
36,540		192,270		36,540	
138,690	3,224,000	138,690	842,000	138,690	1,132,000
	2,404,000		134,000		425,000
	820,000		708,000		707,000
	155,730				155,730
	976,000				868,000
	3,045,000		507,000		953,000
	2,069,000		201,000		85,000

One other point is worthy of comment. There is no apparent pattern of conformity between the ranking of programs in terms of the preliminary locational-advantage figures and the ranking in terms of net overall disadvantages. For example, Nylon J, which has the *largest* preliminary locational-advantage figure, has the *highest* net overall disadvantage.

NYLON SALT PRODUCTION COSTS: MAINLAND OPERATIONS ACHIEVING MODERATE SCALE ECONOMIES

It must be recognized that mainland operations with output scales of the magnitude assumed for "maximum" scale economies are not strictly inevitable. Technological and marketing considera-

TABLE 21 (Cont'd)
SUMMARY OF ADVANTAGES AND DISADVANTAGES
(IN $/YR)

			Nylon F		Nylon G	
Mainland Scale Economies	Maximum	Preliminary Net Locational Advantage:		554,000		927,000
		Process and/or Scale Disadvantages: Nylon Salt Ammonium Nitrate Urea	4,433,290 264,480 258,750	4,957,000	3,088,995 420,210 258,750	3,768,000
		Net Overall Advantage: Disadvantage		4,403,000		2,841,000
	Moderate	Preliminary Net Locational Advantage:		554,000		927,000
		Process and/or Scale Disadvantages: Nylon Salt Ammonium Nitrate Urea	1,347,945 36,540 138,690	1,523,000	3,650 192,270 138,690	335,000
		Net Overall Advantage: Disadvantage:		969,000		592,000
	Minimum	Preliminary Net Locational Advantage		554,000		927,000
		Scale Advantage: Ammonium Nitrate		155,730 710,000		
		Process and/or Scale Disadvantages: Nylon Salt		1,344,000		
		Net Overall Advantage: Disadvantage:		634,000		927,000

tions may continue effectively to limit the scale of individual nylon fiber operations to 35 to 50 MM lb/yr. Furthermore, if transport costs and integration economies dictate that nylon salt activities be regionally adjacent to the fiber operations, at least some nylon salt plants will continue to operate at much less than a capacity of 120 MM lb/yr.

Although ammonia production in the United States is generally large scale, there are strong forces tending to pull ammonia-production activities close to consuming areas.[7] Since urea, ammonium nitrate, and other fertilizer components are often closely integrated with ammonia production, the factors leading to wide-

---
[7] Isard and Schooler, *op. cit.*, p. 38.

DIFFERENTIAL-PROFITABILITY: NYLON

TABLE 21 (Cont'd)
INDIVIDUAL PUERTO RICO NYLON PROGRAMS, BY MAINLAND SCALE ECONOMIES ASSUMPTION

Nylon H		Nylon J		Nylon K	
	957,000		1,087,000		485,000
3,935,065		6,933,905		3,284,105	
184,440		391,230		420,210	
258,750	4,378,000	239,250	7,564,000	258,750	4,633,000
	3,421,000		6,477,000		4,148,000
	957,000		1,087,000		485,000
849,720		3,848,560		738,760	
(adv:)43,000		179,010		192,270	
138,690	988,000	128,238	4,156,000	138,690	1,070,000
	12,000		3,069,000		585,000
	957,000		1,087,000		485,000
	235,770				
	1,193,000				
	846,000		3,845,000		735,000
	347,000		2,758,000		250,000

spread production locations for ammonia will also tend to keep fertilizer component production relatively nonconcentrated and perhaps of more modest scale than 200 MM lb/yr capacity. It thus becomes relevant to consider the magnitude of the differential-profitability correction appropriate for Puerto Rico nylon programs under the assumption that comparable mainland operations will enjoy "moderate" scale economies.

Choice of Output Scales for Specific Activities

The two columns next to the last of Table 19 list the output scales chosen to fit the assumption of "moderate" scale economies in mainland operations. As before, activities outside the complex (e.g., limestone, butadiene, cyclohexane) are not listed. The cost of such products is assumed to be the same for a small or for a large nylon operation. Likewise, scales are not listed for by-product commodities, whose values tend to be set by their opportunity costs. There are of course many possible choices of actual output scales which could be represented as appropriate for moderate scale economies. The scales that appear in Table 19 are intended to be a rough approximation of those that are currently common (if not completely typical) in the United States.

The choice of two possible scales for nitric acid requires comment. It was felt that, if a nylon salt process of the type required in Nylon H or L, with its large nitric acid requirements, were to prove most efficient, then a nitric acid scale of 140 MM lb/yr would certainly become "common." However, if some other process proves most efficient, a smaller scale of nitric acid, 100 MM lb/yr, is more appropriate to moderate scale economies.

Estimated Production Cost, by Process

Table 22 summarizes the estimated production cost of nylon salt by the fourteen different process combinations investigated. The estimates are achieved by using the tables in Appendix C and the assumed utility and capital-cost rates in the manner already described in the discussion of Table 18. A scan of the final total row of Table 22 reveals that the Nylon-G process has the lowest cost. Thus, when the assumption of the magnitude of scale economies is changed from "maximum" to "moderate," the lowest-cost process for nylon salt shifts from that of Nylon D to that of Nylon G. This shift takes place because D requires more labor man-hours than G, and, as the scale of the nylon salt plant decreases, labor costs, direct and indirect, form a larger relative proportion of total costs.

DIFFERENTIAL-PROFITABILITY CORRECTIONS: MODERATE MAINLAND SCALE ECONOMIES

By comparison of the costs in Table 18 with the cost of nylon salt process G in Table 22, the scale and process disadvantage of each Puerto Rico program in nylon salt production can be calculated. Then, by use of the tables in Appendix C on ammonia, nitric acid, ammonium nitrate, and urea, plus the comparative

DIFFERENTIAL-PROFITABILITY: NYLON

scales of the relevant columns of Table 19, the correction applicable to each program's production of fertilizer components can be computed.

Nylon H requires different treatment from the others with respect to the correction for ammonium nitrate. Table 19 indicates that the scale required for nitric acid production in Puerto Rico program Nylon H is 140 MM lb/yr, whereas the scale assumed for nitric acid in a moderate-scale mainland operation of the Nylon-G type (the lowest-cost nylon process) is only 100 MM lb/yr. Though it may seem a bit incongruous to assign one of the Puerto Rico programs a scale-*advantage* correction relative to a mainland operation, this must be done for the ammonium nitrate output (because of its nitric acid input) of Nylon H.

Evaluation of Results

The middle section of Table 21 contains a summary of each program's total differential-profitability correction, plus the amount of its net overall advantage or disadvantage relative to a mainland operation, using the most efficient processes carried on at scales yielding moderate scale economies. This time two of the Puerto Rico programs, Nylon G and Nylon H, emerge with positive overall *advantage* figures. It is not too surprising to note that Nylon G, which utilizes the nylon salt process estimated to have the lowest mainland production costs in both small- and moderate-scale operations, achieves by far the most favorable position of all the Puerto Rico programs compared to a mainland operation. For the other programs, the disadvantage of the higher cost nylon processes remains of considerable magnitude. Such disadvantage when added to the relatively small disadvantage due to fertilizer component production, results in overall disadvantage (save for Nylon H).

NYLON SALT PRODUCTION COSTS: MAINLAND OPERATIONS ACHIEVING MINIMUM SCALE ECONOMIES

It is conceivable that, in the future, mainland conditions of technology, transport, competition, and markets may lead to the production of nylon and of other synthetic fibers in closely integrated operations similar in type and scale to the Puerto Rico programs

INDUSTRIAL COMPLEX ANALYSIS

TABLE 22
PRODUCTION COST OF NYLON SALT

Programs:		A	B	C	D	E
Plant Size: (MM lb/yr)		36.5	36.5	36.5	36.5	36.5
Plant Investment: (MM $)		28	28	29.5	28	28
Raw Materials:						
85% Cyclohexane,	@ 2.628¢/lb	2.865	2.129		2.865	2.129
Furfural,	@ 5.256¢/lb	4.100	4.100	4.100		
Butadiene,	@ 6.566¢/lb				2.561	2.561
Benzene,	@ 2.320¢/lb			1.508		
HCl (100%),	@ 3.223¢/lb	1.515	1.515	1.515		
Sodium Cyanide,	@ 5.150¢/lb	2.626	2.626	2.626		
Hydrogen,	@ 0.900¢/lb	0.036	0.036	0.081	0.045	0.045
Ammonia,	@ 1.880¢/lb					
Sulfuric Acid,	@ 0.683¢/lb					
Hydrogen Cyanide,	@ 3.430¢/lb				0.995	0.995
Limestone,	@ 0.200¢/lb				0.106	0.106
Chlorine,	@ 1.589¢/lb				0.763	0.763
Nitric Acid,	@ 1.959¢/lb					
	@ 2.080¢/lb		1.664	1.664		1.664
Utilities:						
Steam,	@ 40.0¢/M lb	1.262	1.226	1.566	1.252	1.216
Cooling Water,	@ 1.5¢/M gal	0.126	0.133	0.218	0.132	0.150
Electric Power,	@ 0.5¢/kwh	0.200	0.235	0.280	0.200	0.235
Fuel,	@ 15.0¢/MMBtu	0.087	0.066	0.066	0.070	0.048
Direct Labor:						
	@ $2.50/mhr	1.662	1.830	1.975	1.802	1.970
Capital and Indirect Costs:						
Supervision		0.166	0.183	0.198	0.180	0.197
Plant Maintenance		3.068	3.068	3.233	3.068	3.068
Equipment and Operating Supplies		0.460	0.460	0.485	0.460	0.460
Payroll Overhead		0.504	0.532	0.568	0.527	0.555
Indirect Production Cost		2.142	2.216	2.356	2.204	2.278
Depreciation		7.670	7.670	8.082	7.670	7.670
Taxes		0.767	0.767	0.808	0.767	0.767
Insurance		0.767	0.767	0.808	0.767	0.767
Interest		3.068	3.068	3.233	3.068	3.068
Total		33.091	34.291	35.370	29.502	30.712
Less By-product Credit					.161	.161
Final Total		33.091	34.291	35.370	29.341	30.551

analyzed in this study. Only the requirements of the common, high-volume industrial chemicals such as sulfuric acid, chlorine, and hydrogen chloride would come from very large-scale plants. Under such circumstances, the mainland operations would enjoy only "minimum" economies of scale relative to the Puerto Rico programs. The set of differential-profitability corrections for nylon programs which would result on the basis of these assumptions are now examined.

DIFFERENTIAL-PROFITABILITY: NYLON 173

TABLE 22
/100 LB, MODERATE SCALE ECONOMIES

F	G	H	J	K	L	M	N	O
36.5	36.5	36.5	36.5	36.5	36.5	36.5	36.5	36.5
29.5	28	28	28	28	29.5	29.5	29.5	29.5
	7.122	5.282						
			8.304	5.121		4.100		
2.561							2.561	
1.508					3.712	2.018	2.018	5.011
			3.030			1.515		
			5.253			2.627		
0.090	0.036	0.036	0.036	0.054	0.144	0.099	0.108	0.189
	0.338	0.338			0.338			0.338
			0.328	0.328				
0.995				1.989			0.995	
0.106				0.214			0.106	
0.763				1.525			0.763	
		3.879			3.879			
1.664								
1.556	1.544	1.454	1.286	1.264	2.299	1.722	1.712	2.683
0.227	0.227	0.253	0.135	0.225	0.381	0.217	0.217	0.385
0.280	0.305	0.395	0.180	0.175	0.510	0.260	0.260	0.465
0.048	0.120	0.068	0.085	0.049	0.068	0.087	0.070	0.120
2.115	1.270	1.435	1.587	1.753	1.612	1.803	1.958	1.460
0.212	0.127	0.144	0.159	0.175	0.161	0.180	0.196	0.146
3.233	3.068	3.068	3.068	3.068	3.233	3.233	3.233	3.233
0.485	0.460	0.460	0.460	0.460	0.485	0.485	0.485	0.485
0.591	0.439	0.467	0.492	0.569	0.508	0.539	0.565	0.483
2.418	1.970	2.043	2.110	2.182	2.196	2.280	2.349	2.130
8.082	7.670	7.670	7.670	7.670	8.082	8.082	8.082	8.082
0.808	0.767	0.767	0.767	0.767	0.808	0.808	0.808	0.808
0.808	0.767	0.767	0.767	0.767	0.808	0.808	0.808	0.808
3.233	3.068	3.068	3.068	3.068	3.233	3.233	3.233	3.233
31.783	29.298	31.594	38.785	31.423	32.457	34.096	30.527	30.059
.161			1.300	1.655			.161	
31.622	29.298	31.594	37.485	29.768	32.457	34.096	30.366	30.059

Activity-Output Scales: Estimated Production Cost by Process

The middle columns of Table 19 present the output scales assumed for the various activities under the minimum scale economies case. Since the scales for the mainland activities are taken to be the same as those in the corresponding Puerto Rico programs except for the basic chemicals mentioned above, four different sets of

TABLE 23
PRODUCTION COST OF NYLON SALT

Programs:		A	B	C	D	E
Plant Size: (MM lb/yr)		36.5	36.5	36.5	36.5	36.5
Plant Investment: (MM $)		28	28	29.5	28	28
Raw Materials:						
85% Cyclohexane,	@ 2.628¢/lb	2.865	2.129		2.865	2.129
Furfural,	@ 5.256¢/lb	4.100	4.100	4.100		
Butadiene,	@ 6.566¢/lb				2.561	2.561
Benzene,	@ 2.320¢/lb			1.508		
HCl(100%),	@ 3.233¢/lb	1.515	1.515	1.515		
Sodium Cyanide,	@ 5.978¢/lb					
	@ 6.739¢/lb	3.437	3.437	3.437		
Hydrogen,	@ 0.900¢/lb	0.036	0.036	0.081	0.045	0.045
Ammonia,	@ 1.933¢/lb					
Sulfuric Acid,	@ 0.683¢/lb					
Hydrogen Cyanide,	@ 4.296¢/lb					
	@ 4.842¢/lb				1.404	1.404
Limestone,	@ 0.200¢/lb				0.106	0.106
Chlorine,	@ 1.589¢/lb				0.763	0.763
Nitric Acid,	@ 1.970¢/lb					
	@ 2.095¢/lb		1.676	1.676		1.676
Utilities:						
Steam,	@ 40.0¢/M lb	1.262	1.226	1.566	1.252	1.216
Cooling Water,	@ 1.5¢/M gal	0.126	0.133	0.218	0.132	0.150
Electric Power,	@ 0.5¢/kwh	0.200	0.235	0.280	0.200	0.235
Fuel,	@ 15.0¢/MMBtu	0.087	0.066	0.066	0.070	0.048
Direct Labor:						
	@ $2.50/mhr	1.662	1.830	1.975	1.802	1.970
Capital and Indirect Costs:						
Supervision		0.166	0.183	0.198	0.180	0.197
Plant Maintenance		3.068	3.068	3.233	3.068	3.068
Equipment and Operating Supplies		0.460	0.460	0.485	0.460	0.460
Payroll Overhead		0.504	0.532	0.568	0.527	0.555
Indirect Production Cost		2.142	2.216	2.356	2.204	2.278
Depreciation		7.670	7.670	8.082	7.670	7.670
Taxes		0.767	0.767	0.808	0.767	0.767
Insurance		0.767	0.767	0.808	0.767	0.767
Interest		3.068	3.068	3.233	3.068	3.068
Total		33.902	35.114	36.193	29.911	31.133
Less By-Product Credit					.161	.161
Final Total		33.902	35.114	36.193	29.750	30.972

output scales are required. In the usual manner, nylon salt production costs can be estimated for the fourteen possible process combinations. They are summarized in Table 23. Process G again achieves lowest cost. Thus the activity scales shown in the first column of Table 19 are adopted as characteristic of the mainland operation. For this reason the Puerto Rico nylon programs B, C, E, F, and H will each exhibit an *advantage* due to ammonium nitrate production because of their larger-scale nitric acid activities.

TABLE 23
100 LB, MINIMUM SCALE ECONOMIES

F	G	H	J	K	L	M	N	O
36.5	36.5	36.5	36.5	36.5	36.5	36.5	36.5	36.5
29.5	28	28	28	28	29.5	29.5	29.5	29.5
	7.122	5.282	8.304			4.100		
2.561				5.121			2.561	
1.508					3.712	2.018	2.018	5.011
			3.030			1.515		
			6.097					
						3.437		
0.090	0.036	0.036	0.036	0.054	0.144	0.099	0.108	0.189
	0.348	0.348			0.348			0.348
			0.328	0.328				
				2.492				
1.404							1.404	
0.106				0.214			0.106	
0.763				1.525			0.763	
			3.901			3.901		
1.676								
1.556	1.544	1.454	1.286	1.264	2.299	1.722	1.712	2.683
0.227	0.227	0.253	0.135	0.225	0.381	0.217	0.217	0.385
0.280	0.305	0.395	0.180	0.175	0.510	0.260	0.260	0.465
0.048	0.120	0.068	0.085	0.049	0.068	0.087	0.070	0.120
2.115	1.270	1.435	1.587	1.753	1.612	1.803	1.958	1.460
0.212	0.127	0.144	0.159	0.175	0.161	0.180	0.196	0.146
3.233	3.068	3.068	3.068	3.068	3.233	3.233	3.233	3.233
0.485	0.460	0.460	0.460	0.460	0.485	0.485	0.485	0.485
0.591	0.439	0.467	0.492	0.569	0.508	0.539	0.565	0.483
2.418	1.970	2.043	2.110	2.182	2.196	2.280	2.349	2.130
8.082	7.670	7.670	7.670	7.670	8.082	8.082	8.082	8.082
0.808	0.767	0.767	0.767	0.767	0.808	0.808	0.808	0.808
0.808	0.767	0.767	0.767	0.767	0.808	0.808	0.808	0.808
3.233	3.068	3.068	3.068	3.068	3.233	3.233	3.233	3.233
32.204	29.308	31.626	39.629	31.926	32.489	34.906	30.936	30.069
.161			1.300	1.655			.161	
32.043	29.308	31.626	38.329	30.271	32.489	34.906	30.775	30.069

DIFFERENTIAL-PROFITABILITY CORRECTIONS AND EVALUATION OF RESULTS: MINIMUM MAINLAND SCALE ECONOMIES

The amounts of the differential-profitability corrections are calculated in the manner already described and are listed, with the amounts of final overall advantage or disadvantage, in the lower section of Table 21. Puerto Rico program Nylon G suffers no scale or process disadvantage at all, and has associated with it the entire amount of the preliminary locational advantage. In spite of the general absence of scale disadvantage, the other programs, with the exception of Nylon D and Nylon H, still suffer net overall disadvantage compared to the mainland programs. The reason, of course, is their use of less efficient nylon salt processes. With respect to programs D and H, however, it is interesting to note that H achieves a larger overall net advantage than D, in spite of the fact that D has less of a nylon salt disadvantage. This result is due partly to H's larger preliminary locational advantage and partly to its scale advantage in nitric acid production.

SUMMARY

For each of the Puerto Rico nylon programs analyzed in this study, differential-profitability corrections are calculated appropriate to the separate assumptions of maximum, moderate, or minimum scale economies to be enjoyed by comparable mainland operations. The analytical procedure consists of determining the set of estimated mainland production costs of the nylon salt and fertilizer components by the lowest-cost processes at the output scales assumed and comparing that set of costs with the sets of estimated production costs of nylon salt and fertilizer components in mainland programs identical with the Puerto Rico programs. The differences for each program multiplied by the outputs required by the program yield the total differential-profitability correction for the program. This amount is then subtracted from the program's preliminary locational-advantage figure in order to secure the amount representing net overall advantage or disadvantage for the program in Puerto Rico.

Under conditions of maximum mainland scale advantage, the assumption of mainland nylon salt plants of much larger scale than those envisaged for Puerto Rico results in large disadvantages for Puerto Rico programs. For any program these disadvantages more than offset any preliminary locational advantage. However, under the moderate and minimum mainland scale assumptions, for which nylon salt scales are taken to be the same in Puerto Rico and on the mainland, two to three Puerto Rico programs achieve net overall advantages. Of these programs, Nylon G emerges in by far the strongest position.

DIFFERENTIAL-PROFITABILITY: NYLON

It should be noted that for at least some companies nylon salt production on the mainland is much more likely in the forseeable future to be of moderate (or minimum) scale than maximum scale. This statement appears on balance to be the most reasonable in view of the locational considerations, distribution of markets, and current growth patterns applicable to nylon production in the United States.

Chapter 9

Differential-Profitability Analysis: Orlon, Dynel, and Dacron

The general procedure for calculating differential-profitability corrections has been explained in considerable detail in the preceding chapter on nylon programs. It is thus possible in the present chapter to treat in a more summary manner the calculation of the corrections applicable to the programs featuring the production of the other synthetic fibers: Orlon, Dacron, and Dynel.

ORLON

In previous chapters of this study, eight possible Puerto Rico production programs for Orlon fiber were examined and their preliminary locational advantage terms evaluated. As in the nylon programs, each Orlon program calls for the production of one final intermediate commodity for the fiber activity and of two final fertilizer commodities. The former is acrylonitrile; the latter are ammonium nitrate and urea. Each of these three items requires a differential-profitability correction.

Acrylonitrile Production Costs: Mainland Operations Identical with Puerto Rico Programs

Table 24 presents a summary of the estimated production costs of acrylonitrile in eight mainland operations, each one identical

TABLE 24

PRODUCTION COST OF ACRYLONITRILE, $/100 LB, ORLON PROGRAMS IDENTICAL TO PUERTO RICO

Program:		A	B	C	D	E	F	H	J
Plant Size: (MM lb/yr)		40	40	40	40	40	40	40	40
Plant Investment: (MM $)		1.810	1.810	1.810	1.810	1.810	1.810	2.638	2.638
Raw Materials:									
Ethylene Oxide,	@ 8.046¢/lb	8.207	8.207	8.207	8.207				
Acetylene,	@ 9.722¢/lb					9.916	9.916		
Hydrogen Cyanide,	@ 5.120¢/lb	2.234							
	@ 4.381¢/lb		2.191	2.191	2.191	2.191	2.191		
	@ 4.296¢/lb							2.781	
	@ 4.214¢/lb								
	@ 4.129¢/lb								2.725
Utilities:									
Steam,	@ 40¢/M lb	0.336	0.336	0.336	0.336	0.336	0.336	0.832	0.832
Cooling Water,	@ 1.5¢/M gal	0.023	0.023	0.023	0.023	0.023	0.023	0.012	0.012
Electric Power,	@ 0.5¢/kwh	0.235	0.235	0.235	0.235	0.235	0.235	0.190	0.190
Fuel,	@ 15¢/MMBtu	0.038	0.038	0.038	0.038	0.038	0.038		
Direct Labor:	@ $2.50/mhr	0.358	0.358	0.358	0.358	0.358	0.358	0.363	0.363
Capital and Indirect Costs:									
Supervision		0.036	0.036	0.036	0.036	0.036	0.036	0.036	0.036
Plant Maintenance		0.181	0.181	0.181	0.181	0.181	0.181	0.264	0.264
Equipment and Operating Supplies		0.027	0.027	0.027	0.027	0.027	0.027	0.040	0.040
Payroll Overhead		0.073	0.073	0.073	0.073	0.073	0.073	0.281	0.281
Indirect Production Cost		0.241	0.241	0.241	0.241	0.241	0.241	0.080	0.080
Depreciation		0.453	0.453	0.453	0.453	0.453	0.453	0.660	0.660
Taxes		0.045	0.045	0.045	0.045	0.045	0.045	0.066	0.066
Insurance		0.045	0.045	0.045	0.045	0.045	0.045	0.066	0.066
Interest		0.181	0.181	0.181	0.181	0.181	0.181	0.264	0.264
	Total	12.713	12.670	12.670	12.670	14.379	14.379	9.314	9.258

180 *INDUSTRIAL COMPLEX ANALYSIS*

TABLE 25
ACTIVITY SCALES (IN MM LB/YR) FOR PUERTO RICO ORLON PROGRAM

	Scales for Puerto Rico					
	Chlorhydrin		Ethylene Oxide Process Oxidation		Acetylene Process	
	Standard Fertilizer (Program F)	Limited Fertilizer (Program E)	Standard Fertilizer (Programs B,D)	Limited Fertilizer (Programs A,C)	Standard Fertilizer (Program J)	Limited Fertilizer (Program H)
Acrylonitrile	40	40	40	40	40	40
Ethylene Oxide	40	40	40	40		
Ethylene	50	50	50	50		
Acetylene					25	25
Chlorine	66	66				
Quicklime	80	80				
Hydrogen Cyanide	20	20	20	20	25	25
Ammonia	100	80	100	60-80	100	60
Nitric Acid	70	60	70	40-60	70	40
Ammonium Nitrate	90	75	90	50-75	90	50
Urea	70	60	70	40-60	70	40

with one Puerto Rico Orlon program. At the present time there are two economically feasible methods of producing acrylonitrile. One employs ethylene oxide as a principal intermediate, while the other uses acetylene. Both methods require hydrogen cyanide as a second major material input. It can be seen that the first six Orlon programs call for the ethylene oxide method, and the last two the acetylene method. Half the programs require the "standard" fertilizer production, while the other half call for more limited amounts.[1]

Choice of Output Scales for Specific Activities

The first six columns of Table 25 show for each type of Orlon program or program group the required output scales for the basic raw materials and intermediate commodities. This information is necessary in order to choose the correct rate at which to cost the raw materials inputs of Table 24. For example, hydrogen cyanide (HCN) costs more per pound in Orlon A than in Orlon B because,

---

[1] See Appendix B for detailed differences in the programs.

DIFFERENTIAL-PROFITABILITY: OTHER FIBERS

TABLE 25
AND FOR COMPARABLE MAINLAND PROGRAMS, BY SCALE ECONOMIES ASSUMPTION

Scales for Minimum Scale Economies			Scales for Moderate Scale Economies			Scales for Maximum Scale Economies		
Ethylene Oxide Process		Acetylene Process	Ethylene Oxide Process		Acetylene Process	Ethylene Oxide Process		Acetylene Process
Chlorhydrin	Oxidation		Chlorhydrin	Oxidation		Chlorhydrin	Oxidation	
40	40	40	50	50	50	100	100	100
40	40		50	50		100	100	
50	50		50	50		200	200	
		25			50			200
66			66			66		
100			100			100		
20	20	25	50	50	50	100	100	100
100	100	100	100	100	100	135	135	135
70	70	70	100	100	100	200	200	200
90	90	90	100	100	100	200	200	200
70	70	70	100	100	100	200	200	200

in A, ammonia is produced on a smaller scale than in B. The appropriate HCN rates can be obtained from Appendix C, Table C-14. On the other hand, the different rates at which ethylene oxide is costed result not from differences of scale, but from differences in the process used to manufacture the oxide. Appendix C, Table C-12 provides the required data on both processes.

The last nine columns of Table 25 present in order the sets of mainland activity-output scales under the respective assumptions of minimum, moderate, and maximum scale economies. The output scales are an important factor in the determination of rates at which raw-materials inputs are charged. The conditions represented by these various assumptions are generally the same as those depicted in the discussion of the nylon programs.

Differential-Profitability Corrections

The remaining information necessary to the calculation of the differential-profitability corrections for acrylonitrile production is found in Table 26. This table presents the estimated mainland production costs of acrylonitrile by the two possible processes

TABLE 26
PRODUCTION COST OF ACRYLONITRILE (ORLON PROGRAMS), $/100 LB, BY SCALE ECONOMIES ASSUMPTION

Scale Economies:		From Ethylene Oxide and Hydrogen Cyanide			From Acetylene and Hydrogen Cyanide		
		Minimum	Moderate	Maximum	Minimum	Moderate	Maximum
Plant Size: (MM lb/yr)		40	50	100	40	50	100
Plant Investment: (MM $)		1.810	2.117	3.438	2.638	3.126	5.294
Raw Materials:							
Ethylene Oxide,	@ 8.046¢/lb	8.027					
	@ 7.661¢/lb		7.814				
	@ 5.872¢/lb			5.989			
Acetylene,	@ 5.120¢/lb				3.379		
	@ 4.095¢/lb					2.703	
	@ 2.850¢/lb						1.881
Hydrogen Cyanide,	@ 4.296¢/lb	2.191					
	@ 4.129¢/lb		1.911				
	@ 3.747¢/lb			1.747	2.725		
	@ 3.425¢/lb					2.473	2.261
Utilities:							
Steam,	@ 40¢/M lb	0.336	0.336	0.336	0.832	0.832	0.832
Cooling Water,	@ 1.5¢/M gal	0.023	0.023	0.023	0.012	0.012	0.012
Electric Power,	@ 0.5¢/kwh	0.235	0.235	0.235	0.190	0.190	0.190
Fuel,	@ 15¢/MMBtu	0.038	0.038	0.038			
Direct Labor:	@ $2.50/mhr	0.358	0.298	0.173	0.363	0.308	0.183
Capital and Indirect Costs:							
Supervision		0.036	0.030	0.017	0.036	0.031	0.018
Plant Maintenance		0.181	0.169	0.138	0.264	0.250	0.212
Equipment and Operating Supplies		0.027	0.025	0.021	0.040	0.038	0.032
Payroll Overhead		0.073	0.062	0.039	0.080	0.070	0.046
Indirect Production Cost		0.241	0.209	0.140	0.281	0.251	0.178
Depreciation		0.453	0.423	0.345	0.660	0.625	0.530
Taxes		0.045	0.042	0.035	0.066	0.063	0.053
Insurance		0.045	0.042	0.035	0.066	0.063	0.053
Interest		0.181	0.169	0.138	0.264	0.250	0.212
Total		12.670	11.826	9.449	9.258	8.159	6.693

DIFFERENTIAL-PROFITABILITY: OTHER FIBERS

under conditions of minimum, moderate, and maximum scale economies. It should be noted that under any one scale-economy assumption, the ethylene oxide or the acetylene input is charged at only one rate. Each such rate represents the cost of the input by the most efficient (lowest-cost) process, as shown in tables for ethylene oxide and for acetylene in Appendix C.

For each of the three assumptions with respect to mainland scale economies, the acetylene process for acrylonitrile proves to be of lowest cost. Thus, the Puerto Rico programs' disadvantages in acrylonitrile production are computed on the basis of mainland costs via the acetylene process. For example, by column 3, Table 24, a mainland program identical with the Puerto Rico program Orlon C produces acrylonitrile at 12.670¢/lb. At the same time, mainland production under conditions of *moderate* scale economies yields acrylonitrile via the acetylene process at 8.159 cents. The difference, namely 4.511¢/lb, when multiplied by 38,330,000 lb, the annual acrylonitrile requirement, yields a Puerto Rico disadvantage of $1,729,066. In similar fashion the acrylonitrile disadvantage of each Puerto Rico Orlon program, under all three sets of mainland scale assumptions, can be calculated. Then each program's disadvantage due to ammonium nitrate and urea can be computed in the manner described in the nylon discussion.

Evaluation of Results

The summary of the total differential-profitability corrections for the Orlon programs, and their net overall advantages or disadvantages, are presented in Table 27. Orlon J emerges in the strongest relative position. This position is due principally to the utilization in Orlon J of the more efficient acetylene-based acrylonitrile process. Additional favorable factors are: (1) This program possesses the largest amount of preliminary locational advantage of all the Orlon programs, and (2) its fertilizer output scales are not "limited" as are those of some of the other programs. None of the ethylene oxide-based programs achieves a positive overall advantage under any of the three scale-economy assumptions. Even when the scale disadvantages of these latter programs are negligible, the process disadvantages are sufficient to wipe out any preliminary locational advantages.

TABLE 27
SUMMARY OF ADVANTAGES AND DISADVANTAGES OF INDIVIDUAL PUERTO RICO ORLON PROGRAMS
(IN $/YR): BY MAINLAND SCALE ECONOMIES ASSUMPTION

	Program:	Orlon A	Orlon B	Orlon C	Orlon D
Maximum	Preliminary Locational Advantage:	589,000	976,000	793,000	957,000
	Process and/or Scale Disadvantages:				
	Acrylonitrile	2,307,466	2,290,984	2,290,984	2,290,984
	Ammonium Nitrate	494,017	420,210	464,463	420,210
	Urea	302,475	258,750	253,361	258,750
		3,104,000	2,970,000	3,009,000	2,970,000
	Net Overall Advantage:				
	Disadvantage:	2,515,000	1,994,000	2,216,000	2,013,000
Moderate	Preliminary Locational Advantage:	589,00	976,000	793,000	957,000
	Process and/or Scale Disadvantages:				
	Acrylonitrile	1,745,548	1,729,066	1,729,066	1,729,066
	Ammonium Nitrate	351,733	171,390	262,890	171,390
	Urea	222,258	117,300	139,709	117,300
		2,320,000	2,018,000	2,132,000	2,018,000
	Net Overall Advantage:				
	Disadvantage:	1,731,000	1,042,000	1,339,000	1,061,000
Minimum	Preliminary Locational Advantage:	589,000	976,000	793,000	957,000
	Process and/or Scale Disadvantages:				
	Acrylonitrile	1,324,302		1,307,820	
	Ammonium Nitrate	253,725	1,308,000	124,045	1,308,000
	Urea	155,737		45,461	
		1,734,000		1,477,000	
	Net Overall Advantage:		332,000		351,000
	Disadvantage:	1,145,000		-684,000	

TABLE 27 (Cont'd)
SUMMARY OF ADVANTAGES AND DISADVANTAGES OF INDIVIDUAL PUERTO RICO ORLON PROGRAMS (IN $/YR): BY MAINLAND SCALE ECONOMIES ASSUMPTION

	Program:	Orlon E	Orlon F	Orlon H	Orlon J
Maximum	Preliminary Locational Advantage:	481,000	618,000	643,000	1,093,000
	Process and/or Scale Disadvantages:				
	Acrylonitrile	2,946,044	2,946,044	1,004,629	983,165
	Ammonium Nitrate	490,625	420,210	427,983	420,210
	Urea	267,619	258,750	262,433	258,750
	Net Overall Advantage:				1,662,000
	Disadvantage:	3,223,000	3,007,000	1,052,000	569,000
Moderate	Preliminary Locational Advantage:	481,000	618,000	643,000	1,093,000
	Process and/or Scale Disadvantages:				
	Acrylonitrile	2,384,126	2,384,126	442,712	421,247
	Ammonium Nitrate	277,699	171,390	304,717	171,390
	Urea	147,571	117,300	192,836	117,300
	Net Overall Advantage:				710,000
	Disadvantage:	2,328,000	2,055,000	297,000	383,000
Minimum	Preliminary Locational Advantage:	481,000	618,000	643,000	1,093,000
	Process and/or Scale Disadvantages:				
	Acrylonitrile	962,879	1,963,000	214,648	
	Ammonium Nitrate	131,032		219,810	
	Urea	48,019		135,121	
	Net Overall Advantage:			570,000	1,093,000
	Disadvantage:	1,661,000	1,345,000	73,000	

DYNEL

The production of Dynel fiber requires two chemical intermediate materials: acrylonitrile and vinyl chloride. Each can be produced from either ethylene or acetylene, both of which are basic hydrocarbon raw materials. Thus, some Dynel programs produce both intermediates via ethylene processes; others produce both via acetylene processes; and still others call for one intermediate via ethylene and the other via acetylene. It is clear that the programs based on a single hydrocarbon raw material tend to possess significant scale advantages over those based on a combination of hydrocarbon sources. For any given requirement of the final intermediates, the former programs require a single ethylene or acetylene unit, while the latter require an ethylene *and* an acetylene unit, both

TABLE 28
PRODUCTION COST OF ACRYLONITRILE, $/100 LB,
DYNEL PROGRAMS IDENTICAL TO PUERTO RICO

Program:		A	B	F
Plant Size: (MM lb/yr)		20	20	20
Plant Investment: (MM $)		1.115	1.115	1.558
Raw Materials:				
Ethylene Oxide,	@ 11.472¢/lb	11.701		
	@ 9.779¢/lb		9.975	
Acetylene,	@ 5.495¢/lb			3.627
Hydrogen Cyanide,	@ 4.842¢/lb	2.469	2.469	3.196
Utilities:				
Steam		0.336	0.336	0.832
Cooling Water		0.023	0.023	0.012
Electric Power		0.235	0.235	0.190
Fuel		0.038	0.038	
Direct Labor:				
	@ $2.50/mhr	0.623	0.623	0.610
Capital and Indirect Costs:				
Supervision		0.062	0.062	0.061
Plant Maintenance		0.223	0.223	0.312
Equipment and Operating Supplies		0.033	0.033	0.047
Payroll Overhead		0.119	0.119	0.124
Indirect Production Cost		0.376	0.376	0.412
Depreciation		0.558	0.558	0.780
Taxes		0.056	0.056	0.078
Insurance		0.056	0.056	0.078
Interest		0.223	0.223	0.312
	Total	17.131	15.405	10.671

DIFFERENTIAL-PROFITABILITY: OTHER FIBERS 187

of smaller scale. For this reason, differential-profitability corrections are calculated only for the three Dynel programs in which both acrylonitrile and vinyl chloride are produced from a single hydrocarbon raw material. Dynel A and Dynel B are based on ethylene (acrylonitrile via ethylene oxide; vinyl chloride via ethylene dichloride). Dynel F is based on acetylene (acrylonitrile from acetylene and HCN, vinyl chloride from acetylene and HCl).

Acrylonitrile and Vinyl Chloride Production Costs: Mainland Operations, by Scale-Economy Assumption

The computational procedure is exactly similar to that for the Orlon programs, except that there are two fiber intermediates for

TABLE 29
PRODUCTION COST OF VINYL CHLORIDE, $/100 LB, DYNEL PROGRAMS IDENTICAL TO PUERTO RICO

Program:		A	B	F
Plant Size: (MM lb/yr)		25	25	25
Plant Investment: (MM $)		2.850	2.850	2.850
Raw Materials:				
Ethylene Dichloride,	@ 6.325¢/lb	10.816		
	@ 7.125¢/lb		12.184	
Acetylene,	@ 5.495¢/lb			2.363
Hydrogen Chloride,	@ 6.323¢/lb			4.363
Utilities:				
Steam,	@ 40¢/M lb	0.079	0.079	0.120
Cooling Water,	@ 1.5¢/Mgal	0.020	0.020	0.090
Electric Power,	@ 0.5¢/kwh	0.007	0.007	0.050
Fuel,	@ 15¢/MMBtu	0.035	0.035	
Direct Labor:				
	@ $2.50/mhr	0.326	0.326	0.238
Capital and Indirect Costs:				
Supervision		0.033	0.033	0.024
Plant Maintenance		0.456	0.456	0.456
Equipment and Operating Supplies		0.068	0.068	0.068
Payroll Overhead		0.088	0.088	0.074
Indirect Production Cost		0.353	0.353	0.314
Depreciation		1.140	1.140	1.140
Taxes		0.114	0.114	0.114
Insurance		0.114	0.114	0.114
Interest		0.456	0.456	0.456
	Total	14.105	15.473	9.984
	Less By-Product HCl	2.030	2.030	
		12.075	13.443	9.984

INDUSTRIAL COMPLEX ANALYSIS

TABLE 30
ACTIVITY SCALES FOR PUERTO RICO DYNEL PROGRAMS AND FOR COMPARABLE

	Scales for Puerto Rico					Scales for Minimum Acrylonitrile		
	Acrylonitrile			Vinyl Chloride				
	Ethylene Oxide Process		Acetylene Process	Via Ethylene Dichloride	Via Acetylene and HCl	Ethylene Oxide Process		Acetylene Process
	Via Chlor. Program A	Via Oxid. Program B	Program F	Programs A,B	Program F	Via Chlor.	Via Oxid.	
Acrylonitrile	20	20	20			20	20	20
Vinyl Chloride				25	25			
Ethylene Oxide	20	20				20	20	
Ethylene Dichloride				40				
Ethylene	30	30		30		30	30	
Hydrogen Cyanide	10	10	10			10	10	10
Acetylene			20		20			20
Chlorine	66			30-66	20	66		
Quicklime	40				100			
Hydrogen Chloride					20			
Ammonia	100	100	100	100		100	100	100
Nitric Acid	70	70	70	70	70	70	70	70
Ammonium Nitrate	90	90	90	90	90	90	90	90
Urea	70	70	70	70	70	70	70	70

which to calculate comparative production costs, instead of one.[2] For mainland operations identical with the Puerto Rico programs, Tables 28 and 29 present estimated production costs for acrylonitrile and vinyl chloride, respectively. Table 30 lists the sets of output scales required by each Puerto Rico program, and by comparable mainland operations under the separate assumptions of minimum, moderate, and maximum scale economies. Under each of the three general mainland scale assumptions, Tables 31 and 32 summarize for acrylonitrile and vinyl chloride, respectively, their esti-

---

[2] Once again, the reader is referred to the preceding chapter on nylon programs for full details on the construction of the tables to follow, and the manner in which production costs are built up.

DIFFERENTIAL-PROFITABILITY: OTHER FIBERS   189

TABLE 30
MAINLAND PROGRAMS, BY SCALE ECONOMIES ASSUMPTION
(IN MM LB/YR)

Scale Economies		Scales for Moderate Scale Economies					Scales for Maximum Scale Economies				
Vinyl Chloride		Acrylonitrile			Vinyl Chloride		Acrylonitrile			Vinyl Chloride	
		Ethylene Oxide Process		Acet-ylene Pro-cess			Ethylene Oxide Process		Acet-ylene Pro-cess		
Via Ethy-Di-chlor-ide	Via Acet-ylene and HCl	Via Chlor.	Via Oxid.		Via Ethy-Di-chlor-ide	Via Acet-ylene and HCl	Via Chlor.	Via Oxid.		Via Ethy-Di-chlor-ide	Via Acet-ylene and HCl
		50	50	50			100	100	100		
25	25				50	50				100	100
		50	50				100	100			
40				50					100		
30		50	50	50			200	200	200		
		50	50	50			100	100	100		
	20			50		50			200		200
66	66	66			66	66	66			66	66
		100					100				
100											100
100	100	100	100	100	100	100	135	135	135	135	
70	70	100	100	100	100	100	200	200	200	200	
90	90	100	100	100	100	100	200	200	200	200	
70	70	100	100	100	100	100	200	200	200	200	

mated mainland production costs, via the ethylene route and via the acetylene route. As before, the rates at which the raw-material commodities are charged can be determined by reference to the appropriate tables in Appendix C.

Differential-Profitability Corrections

Tables 31 and 32 show that under all scale assumptions the acetylene processes are the cheaper for both acrylonitrile and vinyl chloride. Therefore the mainland costs by the acetylene processes are used as the basis for the final calculation of differential-profitability corrections stemming from the production of acrylonitrile and vinyl chloride. The corrections due to ammonium

TABLE 31
PRODUCTION COST OF ACRYLONITRILE (DYNEL PROGRAMS), $/100 LB, BY SCALE ECONOMIES ASSUMPTION

Scale Economies:		From Ethylene Oxide and Hydrogen Cyanide			From Acetylene and Hydrogen Cyanide		
		Minimum	Moderate	Maximum	Minimum	Moderate	Maximum
Plant Size: (MM lb/yr)		20	50	100	20	50	100
Plant Investment: (MM $)		1.115	2.117	3.438	1.558	3.126	5.294
Raw Materials:							
Ethylene Oxide,	@ 9.779¢/lb	9.975					
	@ 7.661¢/lb		7.814				
	@ 5.872¢/lb			5.989			
Acetylene,	@ 5.495¢/lb				3.627		
	@ 4.095¢/lb					2.703	
	@ 2.850¢/lb						1.881
Hydrogen Cyanide,	@ 4.842¢/lb	2.469					
	@ 3.747¢/lb		1.911		3.196		
	@ 3.425¢/lb			1.747		2.473	2.261
Utilities:							
Steam,	@ 40¢/M lb	0.336	0.336	0.336	0.832	0.832	0.832
Cooling Water,	@ 1.5¢/M gal	0.023	0.023	0.023	0.012	0.012	0.012
Electric Power,	@ 0.5¢/kwh	0.235	0.235	0.235	0.190	0.190	0.190
Fuel,	@ 15¢/MMBtu	0.038	0.038	0.038			
Direct Labor:	@ $2.50/mhr	0.623	0.298	0.173	0.610	0.308	0.183
Capital and Indirect Costs:							
Supervision		0.062	0.030	0.017	0.061	0.031	0.018
Plant Maintenance		0.223	0.169	0.138	0.312	0.250	0.212
Equipment and Operating Supplies		0.033	0.025	0.021	0.047	0.038	0.032
Payroll Overhead		0.119	0.062	0.039	0.124	0.070	0.046
Indirect Production Cost		0.376	0.209	0.140	0.412	0.251	0.178
Depreciation		0.558	0.423	0.345	0.780	0.625	0.530
Taxes		0.056	0.042	0.035	0.078	0.063	0.053
Insurance		0.056	0.042	0.035	0.078	0.063	0.053
Interest		0.223	0.169	0.138	0.312	0.250	0.212
	Total	15.405	11.826	9.449	10.671	8.159	6.693

TABLE 32
PRODUCTION COST OF VINYL CHLORIDE (DYNEL PROGRAMS), $/100 LB, BY SCALE ECONOMIES ASSUMPTION

		From Ethylene Dichloride			From Acetylene and Hydrogen Chloride		
Scale Economies:		Minimum	Moderate	Maximum	Minimum	Moderate	Maximum
Plant Size: (MM lb/yr)		25	50	100	25	50	100
Plant Investment: (MM $)		2.850	4.730	7.740	2.850	4.730	7.740
Raw Materials:							
Ethylene Dichloride,	@ 6.325¢/lb	10.816					
	@ 5.968¢/lb		10.205				
	@ 4.996¢/lb			8.543			
Acetylene,	@ 5.495¢/lb				2.363		
	@ 4.095¢/lb					1.761	
	@ 2.850¢/lb						1.226
Hydrogen Chloride,	@ 3.223¢/lb				2.224	2.224	2.224
Utilities:							
Steam,	@ 40¢/M lb	0.079	0.079	0.079	0.120	0.120	0.120
Cooling Water,	@ 1.5¢/M gal	0.020	0.020	0.020	0.090	0.090	0.090
Electric Power,	@ 0.5¢/kwh	0.007	0.007	0.007	0.050	0.050	0.050
Fuel,	@ 15¢/MMBtu	0.035	0.035	0.035			
Direct Labor:	@ $2.50/mhr	0.326	0.188	0.108	0.238	0.137	0.079
Capital and Indirect Costs:							
Supervision		0.033	0.019	0.011	0.024	0.014	0.008
Plant Maintenance		0.456	0.378	0.310	0.456	0.378	0.310
Equipment and Operating Supplies		0.068	0.057	0.047	0.068	0.057	0.047
Payroll Overhead		0.088	0.059	0.041	0.074	0.051	0.036
Indirect Production Cost		0.353	0.257	0.190	0.314	0.234	0.178
Depreciation		1.140	0.945	0.775	1.140	0.945	0.775
Taxes		0.114	0.095	0.078	0.114	0.095	0.078
Insurance		0.114	0.095	0.078	0.114	0.095	0.078
Interest		0.456	0.378	0.310	0.456	0.378	0.310
Total		14.105	12.817	10.632	7.845	6.629	5.609
Less By-product HCl @ 3.223¢/lb		2.030	2.030	2.030			
Final Total		12.075	10.787	8.602	7.845	6.629	5.609

nitrate and urea are computed by use of appropriate tables in Appendix C, in the manner already described for nylon and Orlon. The summary of all corrections and the final overall net advantage or disadvantage figures for Puerto Rico Dynel programs appear in Table 33.

Evaluation of Results

In view of the significantly higher estimated production costs of acrylonitrile and vinyl chloride via the ethylene route, it is not surprising that the Dynel A and Dynel B programs are subject to net overall disadvantages even under conditions of minimum mainland scale economies. In fact, under the assumption of maximum mainland scale economies, as much as half the amounts of the differential-profitability corrections applicable to Dynel A and Dynel B can be shown to result from the use of the higher-cost processes. On the other hand, Dynel F utilizing the acetylene route to acrylonitrile and vinyl chloride is subject to only a modest overall disadvantage under conditions of moderate mainland scale economies. Under the more favorable assumption of minimum mainland scale economies, Dynel F emerges with a substantial net overall advantage.

DACRON

The production of Dacron fiber also requires two chemical intermediate materials. They are ethylene glycol and dimethyl terephthalate. For each of these intermediates, two economically feasible production processes are possible. Thus there are four process combinations which can be utilized. These combinations constitute the core of the four Puerto Rico Dacron programs.

Ethylene Glycol and Dimethyl Terephthalate Production Costs: Mainland Operations, by Scale-Economy Assumption

Table 34 indicates the activity-output scales required for the operation of the four Puerto Rico Dacron programs, as well as the scales required by comparable mainland operations under the separate assumptions of minimum, moderate, and maximum scale economies. For each scale-economy assumption, Table 35 summarizes the estimated mainland production costs of ethylene glycol

TABLE 33
SUMMARY OF ADVANTAGES AND DISADVANTAGES OF INDIVIDUAL PUERTO RICO DYNEL PROGRAMS (IN $/YR): BY MAINLAND SCALE ECONOMIES ASSUMPTION

	Program:	Dynel A	Dynel B	Dynel F
Maximum	Preliminary Locational Advantage:	696,000	898,000	1,173,000
	Process and/or Scale Disadvantages:			
	Vinyl Chloride	1,487,180	1,801,820	1,006,250
	Acrylonitrile	1,600,145	1,335,550	609,827
	Ammonium Nitrate	420,210	420,210	420,210
	Urea	258,750	258,750	258,750
		3,766,000	3,816,000	2,295,000
	Net Overall Advantage:			
	Disadvantage:	3,070,000	2,918,000	1,122,000
Moderate	Preliminary Locational Advantage:	696,000	898,000	1,173,000
	Process and/or Scale Disadvantages:			
	Vinyl Chloride	1,252,580	1,567,220	771,650
	Acrylonitrile	1,375,408	1,110,812	385,090
	Ammonium Nitrate	171,390	171,390	171,390
	Urea	117,300	117,300	117,300
		2,917,000	2,967,000	1,445,000
	Net Overall Advantage:			
	Disadvantage:	2,220,000	2,069,000	272,000
Minimum	Preliminary Locational Advantage:	696,000	898,000	1,173,000
	Process and/or Scale Disadvantages:			
	Vinyl Chloride	972,900	1,287,540	491,970
	Acrylonitrile	990,318	725,722	
		1,963,000	2,013,000	681,000
	Net Overall Advantage:			491,970
	Disadvantage:	1,267,000	1,115,000	

TABLE 34
ACTIVITY SCALES FOR PUERTO RICO DACRON PROGRAMS AND FOR COMPARABLE

	Capacities for Puerto Rico				Capacities Scale	
	Ethylene Glycol		Dimethyl Terephthalate		Ethylene Glycol	
	Via Chlor. Programs B,D	Via Oxid. Programs A,C	Via Air Oxid. Programs A,B	Via Nitric Acid Oxid. Programs C,D	Via Chlor.	Via Oxid.
Ethylene Glycol	12	12			12	12
Dimethyl Terephthalate			40	40		
Ethylene	10	10			10	10
Chlorine	20				66	
Quicklime	20				100	
Ammonia	100	100	100	100	100	100
Nitric Acid	70-100	70-100	70	100	70-100	70-100
Ammonium Nitrate	90	90	90	90	90	90
Urea	70	70	70	70	70	70

by the chlorhydrin process and by the oxidation process. Table 36 presents similar data for dimethyl terephthalate via air oxidation and via nitric acid oxidation. As indicated by the footnotes to the minimum scale-economy columns in both tables, the data of these columns also pertain to mainland operations identical with the Puerto Rico programs.

Examination of Tables 35 and 36 indicates that, whatever the assumption regarding mainland scale economies, it is cheaper to produce ethylene glycol via the oxidation process, and dimethyl terephthalate by air oxidation of paraxylene. Production costs by these processes thus form the basis for the calculation of those differential-profitability corrections stemming from the production of the final Dacron intermediates. The computing procedure is the same as has been described for the nylon programs. The same procedure is also followed in the calculation of corrections stemming from the production of ammonium nitrate and urea.

All the differential-profitability corrections are summarized in Table 37. They appear together with the resulting net overall advantage or disadvantage figures for the Puerto Rico programs under each of the three mainland scale assumptions.

TABLE 34
MAINLAND PROGRAMS: BY SCALE ECONOMY ASSUMPTION
(IN MM LB/YR)

for Minimum Economies		Capacities for Moderate Scale Economies				Capacities for Maximum Scale Economies			
Dimethyl Terephthalate		Ethylene Glycol		Dimethyl Terephthalate		Ethylene Glycol		Dimethyl Terephthalate	
Via Air Oxid.	Via Nitric Acid Oxid.	Via Chlor.	Via Oxid.	Via Air Oxid.	Via Nitric Acid Oxid.	Via Chlor.	Via Oxid.	Via Air Oxid.	Via Nitric Acid Oxid.
		50	50			100	100		
40	40			50	50			100	100
		50	50			200	200		
		66				66			
		100				100			
100	100	100	100	100	100	135	135	135	135
70	100	100	100	100	100	200	200	200	200
90	90	100	100	100	100	200	200	200	200
70	70	100	100	100	100	200	200	200	200

Evaluation of Results

None of the four Puerto Rico Dacron programs emerges with a net overall advantage when compared with a mainland operation of maximum scale. But, because all four programs have substantial preliminary locational advantages in Puerto Rico, they show up better under the more favorable assumptions concerning mainland scale economies. One, Dacron A, achieves a net locational advantage in Puerto Rico under conditions of moderate mainland scale economies. And Dacron B and Dacron C each achieves a net advantage under conditions of minimum mainland scale economies. Dacron A holds the most favorable position, because it embodies the lower-cost processes for both ethylene glycol and dimethyl terephthalate.

It should be observed that, in every one of the four Dacron programs, the activity that runs at greatest scale disadvantage when compared with maximum-scale mainland operations is the production of DMT. This finding suggests that a "reduced" Dacron program—in particular a reduced Dacron A program—based on DMT import can compare favorably even with maximum-scale mainland operations. This hypothesis is further examined in the following chapter.

196 INDUSTRIAL COMPLEX ANALYSIS

TABLE 35
PRODUCTION COST OF ETHYLENE GLYCOL (DACRON PROGRAMS), $/100 LB, BY SCALE ECONOMIES ASSUMPTION

Scale Economies:	Chlorhydrin Process			Oxidation Process		
	Minimum[1]	Moderate	Maximum	Minimum[2]	Moderate	Maximum
Plant Size: (MM lb/yr)	12	50	100	12	50	100
Plant Investment: (MM $)	2.305	5.637	8.676	2.520	6.154	9.490
Raw Materials:						
Ethylene, @ 4.4¢/lb	3.388			3.652		
@ 2.9¢/lb		2.233			2.407	
@ 2.2¢/lb			1.694			1.826
Chlorine, @ 3.566¢/lb	(1) 6.098					
@ 1.59¢/lb	(2) 2.719	2.719	2.719			
Quicklime, @ 0.989¢/lb	(1) 1.345					
@ 0.570¢/lb	(2) 0.775	0.775	0.775			
Caustic Soda, @ 2.7¢/lb	0.046	0.046	0.046			
Sulfuric Acid, @ 0.95¢/lb	0.017	0.017	0.017			
Catalyst, @ $1.25/lb				0.288	0.288	0.288
Utilities:						
Steam, @ 40¢/M lb	0.452	0.452	0.452	0.414	0.414	0.414
Cooling Water, @ 1.5¢/Mgal	0.081	0.081	0.081	0.090	0.090	0.090
Electric Power, @ 0.5¢/kwh	0.050	0.050	0.050	0.040	0.040	0.040
Fuel, @ 15¢/MMBtu				0.030	0.030	0.030
Direct Labor:						
@ $2.50/mhr	0.905	0.294	0.170	0.925	0.305	0.178
Capital and Indirect Costs:						
Supervision	0.091	0.029	0.017	0.093	0.031	0.018
Plant Maintenance	0.768	0.451	0.347	0.840	0.492	0.380
Equipment and Operating Supplies	0.115	0.068	0.052	0.126	0.074	0.057
Payroll Overhead	0.206	0.082	0.054	0.216	0.087	0.058
Indirect Production Cost	0.752	0.337	0.234	0.794	0.361	0.253
Depreciation	1.920	1.127	0.868	2.100	1.230	0.949
Taxes	0.192	0.113	0.087	0.210	0.123	0.095
Insurance	0.192	0.113	0.087	0.210	0.123	0.095
Interest	0.768	0.451	0.347	0.840	0.492	0.380
Total	(1)17.386 (2)13.437	9.438	8.097	10.868	6.587	5.151
Less By-product Credit	1.610	1.610	1.610			
Final Total	(1)15.776 (2)11.827	7.828	6.487	10.868	6.587	5.151

[1] Amounts in this column preceded by the symbol (1) represent only costs in mainland operations identical to the Puerto Rico programs, Dacron B and Dacron D. Amounts preceded by the symbol (2) represent only costs in mainland operations under conditions of minimum scale economies. All the other amounts apply to both these types of mainland operations.

[2] All cost amounts in this column apply also to mainland operations identical to the Puerto Rico programs, Dacron A and Dacron C.

TABLE 36
PRODUCTION COST OF DIMETHYL TEREPHTHALATE (DACRON PROGRAMS), $/100 LB, BY SCALE ECONOMIES ASSUMPTION

Scale Economies:		Via Air Oxidation of Paraxylene			Via Nitric Acid Oxidation of Paraxylene		
		Minimum[1]	Moderate	Maximum	Minimum[2]	Moderate	Maximum
Plant Size: (MM lb/yr)		40	50	100	40	50	100
Plant Investment: (MM $)		10.750	11.700	18.625	10.750	11.700	18.625
Raw Materials:							
Paraxylene,	@ 4.0¢/lb	2.720	2.720	2.720	2.384	2.384	2.384
Methanol,	@ 2.7¢/lb	0.189	0.189	0.189	0.189	0.189	0.189
Nitric Acid,	@ 2.095¢/lb				2.472	2.472	
	@ 1.849¢/lb						2.182
Sulfuric Acid,	@ 0.68¢/lb				0.096	0.096	0.096
Utilities:							
Steam,	@ 40¢/M lb	-0.122	-0.122	-0.122	0.379	0.379	0.379
Cooling Water,	@ 1.5¢/M gal	0.039	0.039	0.039	0.069	0.069	0.069
Electric Power,	@ 0.5¢/kwh	0.260	0.260	0.260	0.130	0.130	0.130
Fuel,	@ 15¢/MMBtu	0.042	0.042	0.042	0.021	0.021	0.021
Direct Labor:	@ $2.50/mhr	0.719	0.600	0.375	0.625	0.525	0.325
Capital and Indirect Costs:							
Supervision		0.072	0.060	0.038	0.063	0.053	0.033
Plant Maintenance		1.075	0.936	0.745	1.075	0.936	0.745
Equipment and Operating Supplies		0.161	0.140	0.112	0.161	0.140	0.112
Payroll Overhead		0.200	0.169	0.118	0.184	0.157	0.110
Indirect Production Cost		0.811	0.694	0.508	0.770	0.662	0.486
Depreciation		2.687	2.340	1.863	2.687	2.340	1.863
Taxes		0.269	0.234	0.186	0.269	0.234	0.186
Insurance		0.269	0.234	0.186	0.269	0.234	0.186
Interest		1.075	0.936	0.745	1.075	0.936	0.745
	Total	10.466	9.471	8.004	12.918	11.957	10.241

[1] Amounts in this column also apply to mainland operations identical to Puerto Rico programs Dacron A and Dacron B.
[2] Amounts in this column also apply to mainland operations identical to Puerto Rico programs Dacron C and Dacron D.

TABLE 37
SUMMARY OF ADVANTAGES AND DISADVANTAGES OF INDIVIDUAL PUERTO RICO DACRON PROGRAMS (IN $/YR): BY MAINLAND SCALE ECONOMIES ASSUMPTION

	Program:	Dacron A	Dacron B	Dacron C	Dacron D
Maximum	Preliminary Locational Advantage:	1,471,000	1,235,000	1,397,000	1,161,000
	Process and/or Scale Disadvantages:				
	Ethylene Glycol	674,034	1,252,687	674,034	1,252,687
	D.M.T.	907,739	907,739	1,811,792	1,811,792
	Ammonium Nitrate	420,210	420,210	264,480	264,480
	Urea	258,750	258,750	258,750	258,750
		2,261,000	2,839,000	3,009,000	3,588,000
	Net Overall Advantage				
	Disadvantage:	790,000	1,604,000	1,612,000	2,427,000
Moderate	Preliminary Locational Advantage:	1,471,000	1,235,000	1,397,000	1,161,000
	Process and/or Scale Disadvantages:				
	Ethylene Glycol	504,730	1,083,383	504,730	1,083,383
	D.M.T.	366,856	366,856	1,270,909	1,270,909
	Ammonium Nitrate	171,390	171,390	15,660	15,660
	Urea	117,300	117,300	117,300	117,300
		1,160,000	1,739,00	1,909,000	2,487,000
	Net Overall Advantage:	311,000			
	Disadvantage:		504,000	512,000	1,326,000
Minimum	Preliminary Locational Advantage:	1,471,000	1,235,000	1,397,000	1,161,000
	Scale <u>Advantage</u>:				
	Ammonium Nitrate			156,000	156,000
	Process and/or Scale Disadvantages:			1,553,000	1,317,000
	Ethylene Glycol				578,653
	D.M.T.			904,000	904,052
	Net Overall Advantage:	1,471,000			1,482,000
	Disadvantage:		656,000	649,000	166,000

Mainland Scale Economies

Chapter 10

Modifications of the Analysis and Overall Conclusions

FULL PROGRAMS

The results of the differential-profitability correction calculations for the full programs clearly indicate that the number of Puerto Rico programs that emerge with an overall net advantage is rather small compared to the number of programs initially selected and investigated. This finding holds, even under the assumptions of moderate or minimum mainland scale economies. In Table 38 are listed the full programs that do enjoy net overall advantages under the assumption of minimum mainland scale economies, and in some cases also under the assumption of moderate mainland scale economies. The amounts of the net overall advantages of these programs, as well as of their preliminary locational advantages, are recorded. (Table 38 also presents certain data on a reduced Dacron A program which should be ignored for the moment.)

As the discussion in preceding sections has indicated, none of the full programs has a net overall advantage compared to mainland operations achieving maximum scale economies. However, for many of the programs considered, the utilization of other than lowest-cost processes (in terms of mainland costs) contributes as much to their overall disadvantages as do their small activity scales. This conclusion is borne out by the fact that the majority of programs suffer a net overall disadvantage, even under conditions of minimum mainland scale economies.

TABLE 38
FULL AND REDUCED PUERTO RICO PROGRAMS WITH NET OVERALL ADVANTAGES

Program	Preliminary Locational Advantage in $M/yr	Overall Advantage, in $M/yr		
		Mainland Scale Economies Assumption		
		Maximum	Moderate	Minimum
Nylon G	927	negative	592	927
H	957	negative	12	347
D	708	negative	negative	201
Orlon J	1093	negative	468	1093
H	643	negative	negative	73
Dynel F	1173	negative	negative	681
Dacron A	1471	negative	311	1471
B	1235	negative	negative	656
C	1397	negative	negative	649
Reduced Dacron A	---	73[a]	633[a]	1426[a]

a Data are approximations.

Table 38 shows that Nylon G, which is based on a *single* principal raw material, cyclohexane, is the most favorable nylon program. It also indicates that Orlon J and Dynel F, both of which are based on a *single* principal raw material, namely acetylene, are the best of the Orlon and Dynel programs, respectively. The Dacron program that is the most favorable is Dacron A; its production calls for (1) ethylene glycol by the direct oxidation of ethylene, and (2) dimethyl terephthalate by the air oxidation of paraxylene.

The figures of Table 38 suggest no significant degree of comparative superiority as among nylon, Orlon, and Dacron programs for Puerto Rico. Of the three most favorable programs, Dacron A enjoys the largest net advantage under conditions of minimum mainland scale economies, but the smallest under conditions of moderate mainland scale economies. Dynel's prospects appear definitely less favorable than those of the other three fibers.

The discussion in Chapter 7 has touched upon several of the many possible forces and conditions that can interact to determine what the general scale of the relevant mainland operations will be.

MODIFICATIONS OF THE ANALYSIS: CONCLUSIONS 201

On the whole it appears that the assumption of *moderate* scale economies for the mainland is the most reasonable, at least for the foreseeable future.

SHORT PROGRAMS: DIFFERENTIAL-PROFITABILITY ANALYSIS

In the first part of this study, the locational-advantage terms for several short programs are calculated in order to compare such programs with full complexes. It is of interest at this juncture to analyze the possible effects of differential-profitability corrections upon the preliminary findings for these short programs.

In Chapter 7 it is pointed out that differential-profitability corrections have been undertaken only for the petrochemical group of activities. For this reason, there are no corrections to be applied to the initially calculated advantage figures for Short Programs A (staple fiber activity alone) or C (refinery 4 alone). The other four short programs are subject to differential-profitability corrections because they include the production of "standard" quantities of fertilizer components.

Short Programs B and D, with no refinery activities, call for production of the basic fertilizer raw material, ammonia, from fuel oil. The amount of ammonia required is 80 MM lb/yr. By reference to Table C-1 in Appendix C, it can be estimated (by interpolation) that the mainland production cost of ammonia from fuel oil is approximately 2.200¢/lb. From this figure, in the usual way, estimated costs of ammonium nitrate and of urea can be calculated. They prove to be 2.712¢/lb and 2.628¢/lb, respectively. The derivation of the corrections applicable under the three separate assumptions of mainland scale economies proceeds exactly as described for the full programs. The scales of the relevant activities under the various mainland scale-economies assumptions are the same as the scales listed for those activities in Table 34, among others. Table 39 lists the corrections applicable to Short Programs B and D, under conditions of maximum, moderate, and minimum mainland scale economies.

Short Programs E and F involve calculations based on a different cost rate for ammonia, since for these programs refinery-gas hydrogen and hydrocarbons would be available as ammonia raw materials. This same situation occurs in most of the full programs. Hence, the corrections for Short Programs E and F on account of urea and

TABLE 39
SCALE AND PROCESS DISADVANTAGES, SHORT PROGRAMS B AND D (IN $/YR)

Activity		Mainland Scale Economies Assumption		
		Maximum	Moderate	Minimum
Ammonium Nitrate				
87 MM lb	@ 0.605 ¢/lb	526,350		
	@ 0.319 ¢/lb		277,530	
	@ 0.122 ¢/lb			106,140
Urea				
69 MM lb	@ 0.530 ¢/lb	365,700		
	@ 0.325 ¢/lb		224,250	
	@ 0.155 ¢/lb			106,950

ammonium nitrate are the same as those previously calculated for several full programs, e.g., Dacron A. For ammonium nitrate, the amounts of the corrections are $420,210/yr and $171,390/yr, for mainland conditions of maximum and moderate scale economies, respectively. The corresponding figures for urea are $258,750 and $117,300. The totals are $679,000/yr for conditions of maximum mainland scale economies, and $289,000/yr for conditions of moderate mainland scale economies. (Both figures are rounded to the nearest thousand.) No corrections are necessary under conditions of minimum scale economies.

ADJUSTED RESULTS: SHORT PROGRAMS

In Table 40 are summarized final overall-advantage figures for all the selected short programs, by mainland scale-economies assumption. Comparison of Tables 38 and 40 indicates that the effects of differential-profitability corrections do not change the relative desirability of short and full programs. The short programs containing fiber and fertilizer activities (as well as the one with fiber activity alone) emerge with greater locational advantages in Puerto Rico than the full programs. In general, this superiority of the best short programs over the best full programs is enhanced as successively greater mainland scale economies are assumed.[1]

[1] E.g., under conditions of minimum mainland scale economies, Short Program D has an advantage in Puerto Rico of $2.074 MM/yr. This amount exceeds the annual

MODIFICATIONS OF THE ANALYSIS: CONCLUSIONS   203

Among the short programs themselves, the differential-profitability corrections improve the position of Programs E and F relative to that of Programs B and D. This change occurs because the corrections take account of the process disadvantage (i.e. ammonia from fuel oil instead of from refinery gas) of Programs B and D.[2] Of course, the relative positions of Short Programs A and C, to which no corrections are applied, are improved by the differential-profitability corrections applied to the other short programs.

CONTINUOUS-FILAMENT FIBER PROGRAMS

Up to this point, all the computations of net overall advantage or disadvantage have applied to Puerto Rico programs in which the principal product is staple fiber. However, a significant change in the general complexion of the results occurs when programs that produce continuous-filament fiber are considered.

As explained in earlier chapters, a change of the final activity of any of the programs from staple fiber to continuous-filament fiber involves the utilization of considerably more textile labor. (It is assumed that the increase would be the same for all four types of fibers.) Aside from the increased textile labor, continuous-filament programs are identical with the staple fiber programs. Therefore, to derive Puerto Rico's overall locational advantage or disadvantage with respect to any continuous-filament program, a constant amount, representing the improvement in Puerto Rico's position due to the increased amount of textile labor, is to be added to the overall advantage or disadvantage term for the corresponding staple fiber program. As indicated in the note to Table 15, the amount of the constant correction for the increased textile labor is $2.704 MM/yr.

---

advantage, $1.471 MM, of the best full program, Dacron A, by $603,000. Under conditions of moderate mainland scale economies, the advantage in Puerto Rico of Short Program D exceeds that of Nylon G, the best full program, by $1.193 MM/yr. And, under conditions of maximum mainland scale economies, the best full program, Orlon J, incurs a net disadvantage in Puerto Rico of $569,000/yr. On the other hand, Short Program A has a net advantage of $1.595 MM/yr. Thus, Short Program A's margin is $2.164 MM.

[2] Under the assumption of minimum mainland scale economies, this process disadvantage of Short Program D, the best short program, actually *reduces* its preliminary margin of advantage over Dacron A, the full program with highest preliminary-locational advantage. (See Tables 32, 38, and 40.) As already noted, however, under the assumptions of successively increasing mainland scale economies, the best short programs achieve steadily increasing margins of advantage over the best full programs.

TABLE 40
OVERALL ADVANTAGES OF SELECTED PUERTO RICO SHORT PROGRAMS (IN M $/YR); BY MAINLAND SCALE ECONOMIES ASSUMPTION

Short Program	Mainland Scale Economies	Based on Fuel Oil Export Price	Based on Fuel Oil Import Price	Based on Fuel Oil Export Price and		
				Usual Price Differences	No Short Haul Disadvantage	2¢/Bbl Discount on crude; No Short Haul Disadvantage
A. Staple Fiber Alone	Maximum Moderate Minimum	1595	1563			
B. Fertilizer Based on Fuel Oil	Maximum Moderate Minimum	-121 269 558	-168 222 511			
C. Refinery #4 Alone	Maximum Moderate Minimum			-704	-300	-112
D. Staple Fiber And Fertilizer	Maximum Moderate Minimum	1474 1864 2153	1395 1785 2074			
E. Refinery #4 and Fertilizer	Maximum Moderate Minimum			-612 -222 67	-208 182 471	-20 370 659
F. Refinery #4, Fertilizer and Staple Fiber	Maximum Moderate Minimum			983 1373 1662	1387 1777 2066	1575 1965 2254

TABLE 41
OVERALL ADVANTAGES OF SELECTED CONTINUOUS FILAMENT PROGRAMS (IN $M/YR): BY MAINLAND SCALE ECONOMIES ASSUMPTION

Program		Mainland Scale Economies		
		Maximum	Moderate	Minimum
Orlon	A	189	973	1,559
	B	710	1,662	2,372
	C	488	1,365	2,020
	D	691	1,643	2,353
	E	-519	376	1,043
	F	-303	649	1,359
	H	1,652	2,407	2,777
	J	2,135	3,087	3,797
Dynel	A	-366	484	1,437
	B	-214	635	1,589
	F	1,582	2,432	3,385
Dacron	A	1,914	3,015	4,175
	B	1,100	2,200	3,360
	C	1,092	2,192	3,353
	D	277	1,378	2,538
Nylon	A	-2,299	1,135	1,469
	B	-2,586	847	1,182
	C	-3,133	300	635
	D	-863	2,570	2,905
	E	-1,155	2,279	2,619
	F	-1,699	1,735	2,070
	G	-137	3,296	3,631
	H	-717	2,716	3,051
	J	-3,773	-365	-54
	K	-1,444	2,119	2,454
Short Programs	A	4,267	4,267	4,267
	B	-168	222	511
	C	-704	-704	-704
	D	4,099	4,489	4,778
	E	-445	-55	234
	F	3,854	4,244	4,533
Reduced Dacron A		2,777[a]	3,337[a]	4,130[a]

[a] Data are approximations.

Table 41 summarizes the net overall advantage or disadvantage figures for the full programs and the short programs, assuming continuous-filament fiber production. (For the moment, again ignore the data in the row for the reduced Dacron A program.) Clearly an additional advantage of $2.704 MM is enough to give most of the full programs involving Orlon, Dynel, and Dacron a positive overall advantage. This advantage, however, is not suf-

206 INDUSTRIAL COMPLEX ANALYSIS

ficient to yield a positive advantage for nylon programs under conditions of maximum mainland scale economies. There is of course no change in the relative positions of the various programs, except for the worsened position of the three short programs which contain no fiber activities.

EFFECTS OF DIFFERENT ASSUMPTIONS ON WAGE-RATE DIFFERENTIALS

In Chapter 6 a preliminary investigation was made of the effects of a possible disappearance of the favorable mainland wage-rate differential on chemical and petroleum labor. It was clearly shown that these effects were to increase the advantage of Puerto Rico operations. These effects still obtain under different mainland scale-economies assumptions, as the reader can perceive by comparing columns 4, 5, and 6 of Table 42 with columns 1, 2, and 3 of the same table. Columns 4, 5, and 6 assume no chemical–petroleum wage-rate differential between Puerto Rico and the mainland.

It was also shown in the discussion of Table 17 that, if the wages of such labor were regionally equalized, the preliminary locational advantages in Puerto Rico for several of the full programs would approach and in some instances surpass those of the more favorable short programs. These results are modified somewhat by the application of differential-profitability corrections. A comparison of columns 4, 5, and 6 of Table 42 with columns 1 and 2 of Table 17 shows that the inclusion of differential-profitability corrections generally increases the relative attractiveness of the short programs, compared with the full programs. It is true that, given minimum mainland economies of scale, the process disadvantage of ammonia from fuel oil reduces the net advantage of Short Program D, the most favorable short program in Puerto Rico, to a figure below that of Dacron A. However, when the successive assumptions of moderate and maximum mainland scale economies are adopted, the annual locational advantage of Short Program D only falls from $3.638 MM to $3.349 MM and $2.959 MM, respectively; that of Dacron A falls from $3.700 MM to $2.540 MM and $1.439 MM, respectively.

The effects of still another wage-rate differential hypothesis were examined in Chapter 6, namely, the hypothesis that the unfavorable chemical–petroleum wage rate differential of Puerto Rico would be only $0.50 (i.e. cut in half), while the island's favorable

MODIFICATIONS OF THE ANALYSIS: CONCLUSIONS 207

TABLE 42
OVERALL ADVANTAGES OF SELECTED STAPLE FIBER PROGRAMS (IN $M/YR), BY MAINLAND SCALE ECONOMIES ASSUMPTION AND UNDER EACH OF THREE ASSUMPTIONS OF WAGE RATE DIFFERENTIALS

Program		CL = + 100[a] TL = - 75[b]			CL = 0[a] TL = - 75[b]			CL = + 50[a] TL = - 37.5[b]		
		Mainland Scale Economies			Mainland Scale Economies			Mainland Scale Economies		
		Max.	Mod.	Min.	Max.	Mod.	Min.	Max.	Mod.	Min.
		(1)	(2)	(3)	(4)	(5)	(6)	(7)	(8)	(9)
Orlon	A	-2515	-1731	-1145	- 219	+ 565	+1151	-3356	-2572	-1986
	B	-1994	-1042	- 332	+ 335	+1287	+1997	-2823	-1871	-1161
	C	-2216	-1339	- 684	+ 59	+ 936	+1591	-3065	-2188	-1523
	D	-2013	-1061	- 351	+ 292	+1244	+1954	-2851	-1899	-1189
	E	-3223	-2328	-1661	- 643	+ 252	+ 919	-3955	-3060	-2393
	F	-3007	-2055	-1345	- 416	+ 536	+1246	-3735	-2783	-2073
	H	-1052	- 297	+ 73	+1127	+1882	+2252	-1938	-1183	- 813
	J	- 569	+ 383	+1093	+1648	+2600	+3310	-1441	- 489	+ 221
Dynel	A	-3070	-2221	-1267	- 490	+ 359	+1313	-3798	-2949	-1995
	B	-2918	-2069	-1115	- 487	+ 362	+1316	-3708	-2859	-1905
	F	-1122	- 272	+ 681	+1268	+2118	+3071	-1927	-1077	- 124
Dacron	A	- 790	+ 311	+1471	+1439	+2540	+3700	-1657	- 556	+ 604
	B	-1604	- 504	+ 656	+ 869	+1969	+3129	-2377	-1277	- 117
	C	-1612	- 512	+ 649	+ 614	+1714	+2875	-2480	-1380	- 219
	D	-2427	-1326	- 166	+ 42	+1143	+2303	-3202	-2101	- 941
Nylon	A	-5003	-1569	-1235	-2379	+1055	+1389	-5718	-2284	-1950
	B	-5290	-1857	-1522	-2632	+ 801	+1136	-5992	-2559	-2224
	C	-5837	-2404	-2069	-3113	+ 320	+ 655	-6514	-3081	-2746
	D	-3567	- 134	+ 201	-1048	+2385	+2720	-4323	- 890	- 555
	E	-3859	- 425	- 85	-1306	+2128	+2468	-4602	-1168	- 828
	F	-4403	- 969	- 634	-1784	+1650	+1985	-5120	-1686	-1351
	G	-2841	+ 592	+ 927	- 566	+2867	+3202	-3691	- 258	+ 77
	H	-3421	+ 12	+ 347	-1100	+2333	+2668	-4253	- 820	- 485
	J	-6477	-3069	-2758	-3812	- 404	- 93	-7177	-3769	-3458
	K	-4148	- 585	- 250	-1607	+1956	+2291	-4895	-1332	- 997
Short	A	+1563	+1563	+1563	+2944	+2944	+2944	+ 367	+ 367	+ 367
Pro-	B	- 168	+ 222	+ 511	+ 15	+ 405	+ 694	- 119	+ 271	+ 560
gram	C	- 704	- 704	-704	- 462	- 462	- 462	- 639	- 639	- 639
	D	+1395	+1785	+2074	+2959	+3349	+3638	+ 248	+ 638	+ 927
	E	- 612	- 222	+ 67	- 187	+ 203	+ 492	- 498	- 108	+ 181
	F	+ 983	+1373	+1662	+2789	+3179	+3468	- 99	+ 291	+ 580

[a] CL is wage rate differential (Puerto Rico minus mainland) on chemical-petroleum labor, in ¢/man-hr.
[b] TL is wage rate differential (Puerto Rico minus mainland) on textile labor, in ¢/man-hr.

textile wage rate differential would be only $0.375 (also cut in half). Columns 7, 8, and 9 of Table 42 record the corrected locational advantages in Puerto Rico of the selected full and short programs under this hypothesis. As indicated in Chapter 6, such a hypothesis *tends* to decrease the advantage of any Puerto Rico program including fiber production. Comparison of columns 7, 8, and 9 with columns 1, 2, and 3 of Table 42 indicates that the same general change in advantage obtains under different mainland scale-economies assumptions.

Under this latter set of wage-rate differences, the effects of differential-profitability corrections upon the relative advantages of short and full programs may be noted. Given minimum mainland economies of scale, Short Program D's margin of annual advantage over Dacron A is only $323,000. (See column 9, Table 42.) But, under conditions of moderate and maximum mainland scale economies, that margin rises to $1.194 MM and $1.905 MM, respectively. (See columns 8, and 7.)

Thus, in general, the greater the magnitude of scale economies assumed possible for mainland operations, the more the differential-profitability corrections increase the relative advantage in Puerto Rico of the most favorable short programs over the most favorable full programs. This tendency holds true whatever assumption is adopted regarding the differences between Puerto Rico and the mainland in wage rates for chemical-petroleum labor and for textile labor. However, this conclusion should not obscure the fact that, under any *one* assumption as to mainland scale economies, reduction of Puerto Rico's wage-rate disadvantage on chemical-petroleum labor would definitely *reduce* the relative disadvantage of including petrochemical activities in a Puerto Rico complex.

APPRAISAL OF POSSIBLE METHODS FOR REDUCING SCALE DISADVANTAGES

Because potential scale disadvantages can significantly limit the attractiveness of many of the full Puerto Rico production programs, it is pertinent to inquire if there are any possibilities of reducing such scale disadvantages. Obviously the short programs that have been discussed furnish one answer. For example, Short Program A includes only fiber activities and thus is not subject to a scale disadvantage under any of the three assumptions as to mainland economies of scale. However, it is of interest to consider the following changes in the full programs which are not of such drastic nature.

Importation of One or More Chemical Intermediates

As has already been intimated, one possibility for reducing scale disadvantage is via import of chemical intermediates. To illustrate, consider the program Dacron A. This program suffers no process disadvantage, but confronts scale disadvantages under the assumptions of maximum and moderate mainland scale economies. If a reduced Dacron A program is established which calls for the im-

MODIFICATIONS OF THE ANALYSIS: CONCLUSIONS 209

port of DMT (dimethyl terephthalate) instead of the production of DMT from imported paraxylene, the differential-profitability correction under conditions of maximum mainland scale economies would be cut by $907,739 (the scale disadvantage on DMT) less $45,070 (the excess of DMT transport cost over paraxylene transport cost), or $862,669. This saving would be enough to give the Dacron A staple fiber program in its *reduced form* a positive overall advantage even under the assumption of maximum mainland economies of scale, and would involve the elimination of only one productive activity of the Puerto Rico complex. Relevant data for the reduced Dacron A program (staple and continuous filament) are recorded in the last row of both tables 38 and 41.[3] The superiority of this program over all full programs is clear-cut. Also, this superiority exists under the various assumptions on wage-rate differentials that have been examined.

The import of ethylene glycol for Dacron A would still further reduce the total scale disadvantage. However, because such import would eliminate two activities (ethylene and ethylene glycol) from the Puerto Rico program, it would not be as desirable a change as the import of DMT. In fact, if both ethylene glycol and DMT were imported, the program would be virtually the same as Short Program F (refinery, fiber, and fertilizer activities).

In a similar manner, other reduced Dacron programs might be considered; but, from the analysis already performed, these programs are likely to remain less favorable for Puerto Rico than a reduced Dacron A program. Also, reduced Dynel programs might be examined. (Reduced nylon or Orlon programs based on the import of nylon salt or acrylonitrile, respectively, would not be considered. Such reduced programs would be almost identical with the short programs already examined.)[4] Reduced Dynel programs would avoid the scale disadvantage on acrylonitrile or vinyl chloride (one of which would be imported), but would incur additional scale disadvantage due to the smaller-scale acetylene or ethylene oxide unit of the reduced program. Also involved would be the elimination of

---

[3] The amounts of the reduced Dacron A program advantages recorded in tables 38 and 41 are approximations. They are based on the advantage figures of Dacron A adjusted for the lessened scale disadvantage of the reduced program. Not taken into account are minor adjustments which result from differences between the full and reduced Dacron A programs in such inputs as chemical–petroleum labor, steam, fuel, and power.

[4] Rigorously speaking, some of the nylon programs that have been designated full programs have already been reduced at their very first stages.

additional intermediate activities such as ethylene or hydrogen cyanide production. For these reasons, reduced dynel programs are not likely to be particularly desirable.

Production of Chemical Intermediates for External Markets

One further general possibility exists for the reduction of the differential-profitability corrections due to scale disadvantage in the Puerto Rico programs. If markets exist or could be developed for the various petrochemical intermediates in addition to internal requirements of the programs, the output scales of such petrochemical activities could be greater, and their corresponding unit-scale disadvantages smaller. Two general possibilities may be envisioned. First, chemical products might be shipped to mainland markets. Second, the industrial development of Puerto Rico and the whole Caribbean area might create rather extensive additional markets for organic industrial chemicals of the type required in the fiber programs.

The first possibility does not appear to offer much potential gain for Puerto Rico programs, particularly under conditions of maximum mainland scale economies. One reason is that, unless the scales of the relevant Puerto Rico chemical activities were increased sufficiently to approach those of the mainland, the total amount of scale disadvantage would remain virtually the same. In fact this amount could even *increase,* although the scale disadvantage per unit output would be lessened. For example, if in the Dacron A program the production of ethylene glycol were increased from 12 MM lb/yr to 50 MM lb/yr, the scale-disadvantage correction per pound, compared to a mainland operation of 100 MM lb/yr, would drop from 5.7 cents to 1.4 cents. However, the total scale-disadvantage correction would *increase* from \$674,000 to \$718,000.

Another unfavorable result of increasing the Puerto Rico petrochemical operations would be an increase of total locational-disadvantage amounts, due to Puerto Rico's disadvantage in fuel and chemical labor costs. Such an increase would tend to offset the decrease in scale disadvantages as the Puerto Rico activities approached mainland scale. Obviously, the effects of these unfavorable aspects of expanding Puerto Rico petrochemical-activity scales would be less significant if typical mainland operations could achieve only moderate scale economies.

The second possibility mentioned above, the growth of Caribbean nonfiber markets for petrochemicals, would appear to be con-

MODIFICATIONS OF THE ANALYSIS: CONCLUSIONS 211

siderably more favorable for an expansion of chemical-activity scales. Although the total scale-disadvantage correction would not decrease until mainland scales were approached, and although the locational-disadvantage amounts due to fuel and labor cost would increase, Puerto Rico might acquire a locational advantage from saving in transport cost in serving nearby parts of the Caribbean markets. With respect to the programs computed in this study, such a locational-advantage term would approximately approach twice the transport cost of the chemical product from the mainland to Puerto Rico.[5]

To illustrate the above point, consider again an increase to 50 MM lb/yr of ethylene glycol production in the Dacron A program. Assuming maximum mainland scale economies, the total scale-disadvantage correction would increase from $674,000 to $718,000. But the total transport cost of shipping 38 MM lb of glycol between the mainland and Puerto Rico is $145,000. Twice this amount is $290,000, which roughly approximates an advantage for the Puerto Rico program, and which would more than offset the increase in the total scale-disadvantage correction. However, as already pointed out, the Puerto Rico program would be subject to increases in those disadvantages due principally to high fuel and chemical labor costs. The amounts of such increases would more than wipe out the savings otherwise achieved by the larger-scale production of ethylene glycol.

Hence, it may be concluded that an expansion of the petrochemical-activity scales in a Puerto Rico program cannot be expected to have a net favorable effect except possibly under conditions of (*a*) an adequate local market for the chemical products, and (*b*) only moderate mainland scale economies, or (*c*) no differential in favor of the mainland in the cost of chemical labor.

OVERALL CONCLUSIONS[6]

In this concluding section the major findings of the several chapters of this study are brought together and reviewed. The first set

---

[5] The double saving would result because the increase of chemical production in the program could reasonably be expected to decrease the volume of some refinery product, e.g., fuel oil, to be shipped to the mainland.

[6] In this chapter no attempt is made to appraise generally the industrial complex technique as an approach to regional analysis. Such appraisal and a general formulation of this technique, partly based upon this case study, are contained in Walter Isard, et. al., *Methods of Regional Analysis,* to be published, chapter 9.

of findings are those developed in Chapter 6. There the locational advantages and disadvantages in Puerto Rico of a selected group of full and short programs were investigated under the assumption of identical Puerto Rico and mainland operations. The data summarized in Table 15 indicated for each of the selected full programs in Puerto Rico: (1) disadvantage for Puerto Rico in terms of nonlabor inputs and outputs, (2) disadvantage in chemical-petroleum labor, (3) considerable advantage in textile labor, (4) *net* locational advantage in terms of all factors considered up to that point. Similarly, the data reported in Table 16 indicated preliminary net locational advantage for each of the selected short programs but one.

Furthermore, the preliminary locational advantage of the best short program (the one involving only fiber and fertilizer activities) was shown to exceed substantially the preliminary locational advantage of the most favorable full program. However, this short program's margin of advantage was considerably smaller when alternative calculations were made on the basis of narrower wage differentials between Puerto Rico and the mainland.

The next set of findings are those reported in Chapters 8 and 9. These findings take into account differential profitabilities arising from process and scale differences between Puerto Rico and mainland complexes. Chapter 8 dealt with the selected full programs involving nylon fiber production. Only two to three of these programs retained positive locational advantage in Puerto Rico after the differential-profitability calculations were made. And even those positive figures depended on the assumption that mainland scale economies would be limited to minimum or moderate amounts. Under the assumption of maximum mainland scale economies—an assumption that is not the most realistic—none of the nylon full programs showed net locational advantage for Puerto Rico.

Chapter 9 extended the differential-profitability analysis to the selected full programs involving Orlon, Dynel, and Dacron. Several of these programs were favorable for Puerto Rico under the assumption of minimum mainland scale economies, but only two, Orlon J and Dacron A, retained such a position when moderate mainland scale economies were assumed. None of the programs retained a net locational advantage in Puerto Rico when compared to a mainland operation achieving maximum scale economies.

The final set of findings are reported in previous sections of this chapter. These findings relate to (1) the effects of differential profitabilities on the short programs, (2) the advantages of con-

MODIFICATIONS OF THE ANALYSIS: CONCLUSIONS 213

tinuous-filament fiber programs, (3) the effects of differential-profitability corrections on the results computed for both full and short programs under alternative sets of wage-rate differentials, and (4) the potential advantages for Puerto Rico of certain reduced full programs.

Two findings that are quite positive for Puerto Rico emerge. One is that a reduced Dacron A program, based on the import of DMT, evidences a net advantage for Puerto Rico even under the assumption of maximum-scale mainland operations. Thus under this assumption the reduced Dacron A program is superior to any of the full programs. A second positive finding with respect to Puerto Rico is that continuous-filament fiber production adds approximately a $2.700 MM/yr advantage to all Puerto Rico programs —full, short, and reduced—which include fiber activities. Thus many of the full programs in Puerto Rico are able to compare favorably with mainland programs, even under the assumption of maximum-scale mainland operations.

It is also noted that consideration of differential profitabilities does not appreciably alter the preliminary patterns of *relative* advantage among the short programs or between the short and the full programs. The computed amounts of locational advantage or disadvantage for the individual programs of course vary markedly in the different situations postulated.

Consideration of the three major sets of findings indicates the conclusion that of the six short programs considered, Short Program A (fiber activities only), Short Program D (fiber and fertilizer activities), and Short Program F (refinery, fiber, and fertilizer activities) are the most favorable for Puerto Rico. Of the twenty-eight full programs considered, Nylon G, Orlon J, and Dacron A are the most favorable for Puerto Rico. In general, the most advantageous short programs achieve greater locational advantages in Puerto Rico than the most favorable full programs. The margins of advantage that the short programs have increase as the scale of mainland operations mounts. An important related conclusion is that, relative to short programs, some of the major disadvantages of full programs can be eliminated by scheduling chemical intermediate imports, i.e., by considering "reduced" programs.

These conclusions cannot be considered final. They must be further qualified. As pointed out in Chapter 2, the calculations of locational advantage performed in this study pertain to only the objective and systematic elements of regional cost and revenue differentials. Yet there are many other elements, tangible and intangible, certain and uncertain, which will undoubtedly come to influence the situation at any point of time. These other elements do affect relative profitabilities and are especially important in a comparison of short and full programs. For example, the corporate tax situation in Puerto Rico tends to improve the relative attractiveness of the full programs. Under certain conditions, new corporations are exempt from corporate income taxes for a limited number of years. In view of the level of mainland corporate income taxes, it is evident that a full Puerto Rico complex would enjoy a considerable tax advantage over a split-location operation including only a short program in Puerto Rico.

To illustrate, a hypothetical comparison may be made of the Dacron A program and Short Program F. The former comprises three more activities than the latter; namely, the production of ethylene, ethylene glycol, and DMT. Under conditions of maximum mainland activity scales, the portion of plant investment applicable to the production of the Dacron A program's requirements of these three commodities would be approximately $9 MM. If an average annual profit, before taxes, of 20% on investment is assumed, the profit attributable to these three petrochemical activities would be $1.8 MM. Assuming mainland corporate income taxes at a rate of 50%, the income-tax exemption in Puerto Rico would increase Dacron A's annual locational advantage $900,000 more than that of Short Program F. The relevant figures of column 1, Table 42, indicate that, before taxes are considered, Dacron A suffers a relative disadvantage of $1.773 MM compared with Short Program F. Thus the tax advantage would reduce by more than one-half this relative disadvantage of the full program in Puerto Rico.[7]

If conditions of moderate mainland activity scales are postulated, the tax advantage improves the full program's position even more. Under such conditions, the portion of plant investment applicable to the production of Dacron A's requirements of ethylene, ethylene glycol, and DMT would be $11.491 MM. Again assuming average profit before taxes of 20% on investment, and a mainland corporate

---

[7] It should be noted that any improvement in locational advantage for Puerto Rico gained by substituting a reduced full program (e.g., reduced Dacron A) for the corresponding regular full program would be offset to some extent by any concurrent reduction in potential income-tax benefits in Puerto Rico. Such reduction would be occasioned by the fact that, in the reduced full program, mainland investment, production, and profits would, for one or more commodities, take the place of investment, production, and profits in Puerto Rico.

MODIFICATIONS OF THE ANALYSIS: CONCLUSIONS 215

income-tax rate of 50%, Puerto Rico's tax exemption situation would increase Dacron A's locational advantage by $1.149 MM more than that of Short Program F. Since the latter program's margin of advantage before taxes are considered is only $1.062 MM (see column 2, Table 42), the tax advantage would give *Dacron A* a margin of $87,000.[8]

As stated in Chapter 2, there is justification for the belief that the tax situation and other factors, although difficult to quantify, tend to establish the full programs as more desirable in Puerto Rico than the short programs. The factors that relate to spatial integration economies and social welfare considerations are of particular importance here. As previously noted, full programs tend to have more desirable effects upon rates of industrialization, changes in entrepreneurial attitude, and investment and saving habits than short programs. Full programs also tend to generate greater spatial integration economies associated with growth of pools of efficient labor and of management skills, and with other factors noted in Chapter 2. Hence the margins of advantage that short programs have over full programs are to be interpreted as minimum magnitudes by which the spatial integration and social advantages of full programs must exceed those of short programs.

In the opinion of the authors, the overall results of this study indicate that, for at least a limited type of development, the advantages of a Puerto Rico location with regard to a relatively fully integrated industrial complex are clear-cut. Exactly how extensive this development should be—under favorable circumstances it could be quite extensive—and what its specific form should take obviously depend on a host of subsidiary factors which fall outside the scope of this study.

---

[8] The effects of such a hypothetical tax-profit situation on the relative attractiveness of Dacron A and Short Program D are even more striking. In addition to the petrochemical activities already enumerated, Short Program D excludes the refinery activities. The investment for a type-4 refinery amounts to roughly $100 MM. Thus the tax advantage on account of profits on refinery investment alone would increase Dacron A's locational advantage approximately $10 MM more than that of Short Program D. This would put Dacron A in a much superior position to Short Program D.

Appendix A

Notes on Refinery and Ethylene-Separation Activities

1. The basic operational schemes for the six refinery activities are given in Table A-1. This table indicates that the prototypes were built up by varying the scale of four basic refinery processes: (1) topping (atmospheric distillation), (2) fluid catalytic cracking, (3) coking, (4) reforming. In the case of refinery 5, heavy crude was assumed to be charged; this changes the topping and coking operations to some extent.

In addition to the four processes just mentioned, gasoline stabilization, gas separation, catalytic polymerization, and the recovery of liquefied petroleum gas (LPG) have been included in the scheme of all refineries. The scales of these latter processes were derived from the scales of the basic processes by a set of special assumptions. Finally, a scheme for ethylene separation was worked out for each refinery.

The sources of the technical data used in estimating inputs and outputs of the refinery and gas-separation activities are listed in a bibliography at the end of this appendix.

2. The schemes of refinery prototypes which were adopted represent a condensed version of a much larger number of prototypes considered in detail by Lindsay (source *S1* in the bibliography). The first two chapters of Lindsay's work give an excellent overall picture of refining technology and the problems of constructing representative refinery prototypes from the very large number of individual refinery processes in use today. Lindsay's presentation

TABLE A-1
BASIC OPERATIONAL SCHEMES OF REFINERY PROTOTYPES

Activity

Basic Refinery Process		#1	#2	#3	#4	#5	#6
#1	Topping of light crude, bbl/day	28,570	28,570	28,570	28,570		
#1A	Topping of heavy crude, "					28,570	28,570
#2	Fluid catalytic cracking, "	10,000	15,000	10,000	15,000	11,698	15,000
#3	Coking of heavy residual obtained from light crude, "			2,857	2,857		
#3A	Coking of heavy residual obtained from heavy crude, "					2,857	
#4	Reforming			5,000	5,000	5,000	5,000

The six activities are based on Lindsay's prototypes (Source S1) as follows:

#1. "III-b", pp. 132-136.

#2. "Alternate III-b", pp. 137-141.

#3. "XI-b", pp. 167-169.

#4. "Alternate XI-b", (Not calculated in detail by Lindsay.)

#5. Not based on a Lindsay prototype.

#6. "Alternate V-b (2)". (Not calculated in detail by Lindsay.)

APPENDIX A

is addressed to nontechnical readers. Detailed technical summaries of current refinery processes are published every year in the engineering periodicals, primarily in *Petroleum Refiner* and the *Oil and Gas Journal*.

In constructing refinery prototypes, Lindsay has selected representative processes from the many available alternatives and has adopted standard yields. He has also eliminated many minor refinery processes (desulfurization, chemical treatment of stocks, blending, etc.) from explicit consideration, on the ground that they do not affect location decisions significantly. In outlining the six refinery activities, the authors have accepted Lindsay's judgments, even though the resulting scheme is not as accurate for a United States–Puerto Rico comparison as for United States mainland comparisons (for which Lindsay's materials were developed). Also, it was necessary to supplement Lindsay's materials since they do not provide necessary detail on refinery gases. In other respects, though more detail would have been desirable, the expenditure of the research resources required for such a purpose was not justified in terms of the general objectives of this study.

3. Table A-2 summarizes the inputs and outputs of the basic refinery processes. Most of the inputs and outputs were taken from Lindsay. The gas outputs were collected from a variety of other sources, since Lindsay does not report detailed gas compositions. The topping of heavy crude and the coking of heavy residual obtained from heavy crude were obtained by adjusting Lindsay's figures on the basis of other sources. Details follow.

Topping of Light Crude

This operation also includes vacuum flashing, which is a feed-preparation method for fluid catalytic cracking. Lindsay (*S1*, p. 132) gives the liquid streams and utilities; for gases, Lindsay gives only the heating value. On the basis of Lindsay's data, which agree in order of magnitude with source *S3* (column 1 of cases I-VI, pp. 127-132), it is evident that the amount of the gases is very small compared with cracking, coking, and re-forming gases. Source *S2* (p. 96) reveals that the composition of topping gases varies a great deal with the crude. Consequently, it was decided to ignore topping gases completely.

Topping of Heavy Crude

Utilities and gases were assumed to be the same as for light crude, but the yields of heavy liquid streams have been increased

220 INDUSTRIAL COMPLEX ANALYSIS

TABLE A-2
INPUTS AND OUTPUTS OF REFINERY OPERATIONS

	Commodities	Units	Topping of Light Crude	Topping of Heavy Crude	Fluid Catalytic Cracking	Coking of Heavy Resid. Obtained from Light Crude	Coking of Heavy Resid. Obtained from Heavy Crude	Reforming
1A	Crude Oil, Light	BBL/Day	-28570					
1B	" Heavy	"		-28570				
2	Gasoline, S.R. + Coked	"	+6286	+3000		+653	+2854	-3000
3	" " Cracked	"			+4497			
4	" " Reformed	"						+4505
5	" " Polymerized	"						
6	Naphtha	"	+2000	+2000				-2000
7	Kerosene	"	+2857	+2172				
8	Diesel Oil	"	+4286	+3257				
9	Gas Oil	"			+4000	+1571	+6698	
10	Cycle Oil	"						
11A	Hvy Resid. Fr. Light Crude	"	+2857			-2857		
11B	" " " Heavy "	"		+12857			-12857	
11C	Light Residual	"	+10000	+5000	-10000			
12	Coke and Carbon	LB/Day				+122220	+812305	
13	L.P.G.	"			+2590	+958	+4179	+21790
14	Hydrogen	"			+38150	+35910	+156855	+11350
15	Methane	"			+18370	+22608	+98755	
16	Ethylene	"			+27880	+31569	+137891	+20100
17	Ethane	"			+88030	+51739	+225987	
18	Propylene	"			+59770	+6723	+29365	+45810
19	Propane	"			+95050	+23171	+101210	
20	Butylenes	"			+37200	+14788	+64594	+78040
21	Butanes	"						
22	Pure Ethylene	"						
23	Pure Ethane	"						
24	Steam	M LB/Day	-1080	-1080	-1000	-357	-1607	-696
25	Power	Kwh/Day	-2784	-2784	-4410	-1372	-6171	-240
26	Fuel	MMBtu/Day	-3086	-3086	-1100	-796	-3574	-1800
	Labor, Direct	MHR/Day	-152	-152	-196	-48	-70	-112

at the expense of light streams. For a discussion of product yields from various crudes, see S8.

Fluid Catalytic Cracking

Utilities and liquid streams are based on *S1*. Gas compositions and amounts are based on *S2* through *S7*. Specifically, the total amount of gas is assumed to be 15 wt % of the feed, and the composition is based on an average calculated from the cited sources.

When the feed input to fluid catalytic cracking is increased above 10,000 bbl per day, Lindsay (*S1*, p. 137) takes the increased feed from the Diesel oil stream and from part of the kerosene stream of the topping operation. Lindsay's procedure has been followed in activities 2 and 6. However, in activities 4 and 5, the gas oil obtained from coking is scheduled for cracking feed before the Diesel oil and kerosene are drawn upon. The reason for this procedure is that Diesel oil and kerosene are more valuable when marketed. This distinction, however, is a fine one. It is entirely compatible with the body of Lindsay's discussion. In the present study it does not affect the results, since the price differences applicable to all liquid streams happen to be identical.

When the scale of the fluid catalytic-cracking operation is increased, all outputs increase in proportion, as do all utility inputs. Labor inputs vary according to a 0.25 labor factor.

Coking of Heavy Residual Obtained from Light Crude

Liquid streams (except gasoline) and utilities are taken from Lindsay (p. 167). Lindsay's gasoline output includes all the butanes and butylenes. These are reported separately in Table A-2 of this appendix. Gas outputs and labor inputs are taken from *S9*.

Coking of Heavy Residual Obtained from Heavy Crude

The inputs and outputs for this process may be derived by adjusting the data which refer to the coking of heavy residual obtained from light crude. Based on correspondence with engineers, the suggested adjustments for a typical heavy feedstock include (a) raising the total gases from 18.68 to 25 wt % of the feed, (b) leaving the composition of the gases the same, (c) raising coke from 13 to 15 wt % of the feed (possibly as high as 20% for very heavy feeds), (d) raising gasoline from 17.3 to 24 wt %. Labor and utilities are taken from Lindsay. All inputs and outputs except labor are increased in proportion to total feedstock. For labor, a 0.25 factor is assumed.

Re-forming

Liquid streams (except gasoline), utilities, and labor are taken from Lindsay's Platforming process (*S1*, p. 124). Gases are based on data from *S11* through *S15*, on the assumption of severe operating conditions. Gasoline is based on *S7*. The data were checked by reference to *S16*. It should be noted that the representation of re-forming by a single input–output scheme is not entirely satisfactory, since there is a wide range of possible processes as well as possible variations in operating conditions.

4. In addition to the basic refinery processes, every refinery is assumed to have gasoline-stabilization and gas-separation plants, a catalytic-polymerization unit, and a processing section for separating light, medium, and heavy gaseous components.

The gas streams shown in Table A–2 among the outputs of the basic refinery processes are assumed to come from the gasoline-stabilization plants attached to the individual processes. Such gas streams contain only part of the heavier gaseous components produced; the rest is retained in the gasoline to bring its vapor pressure up to specifications.

The gases from each process form a mixed stream. The gases from cracking and coking are assumed to be united before further processing, while the re-former gases (which are rich in hydrogen) are assumed to be handled separately.

The platformer gases contain no ethylene, propylene, or butylenes. In Figure A–1, their components are denoted by the symbols $P \times 1, 1, 0, 1, 0, 1, 0, 1$. The numbers refer, in turn, to hydrogen, methane, ethylene, ethane, propylene, propane, butylenes and butanes. The numbers 1 refer to the fact that each component is present at unit level; the unit level is defined as the amount of each component in the platformer stream. In the case of missing components, however, the figure 1 is replaced by 0. As depicted, the platformer stream goes to a separation unit (Light Sep.) where the hydrogen and methane content are removed, together with 8% of the original amount of ethane. The resulting streams are: (1) a light stream (Plat) whose components are $P \times 1, 1, 0, 0.08, 0, 0, 0, 0$, i.e., all the hydrogen and methane of the original platformer gas, and 8% of the ethane; and (2) a heavier stream, $P \times 0, 0, 0, 0.92, 0, 1, 0, 1$, which contains all the propane and butane and 92% of the ethane in the original platformer stream. The heavier stream goes to an LPG separator where 89.5% of the propane and all of the butane is removed as LPG; the remainder of

APPENDIX A 223

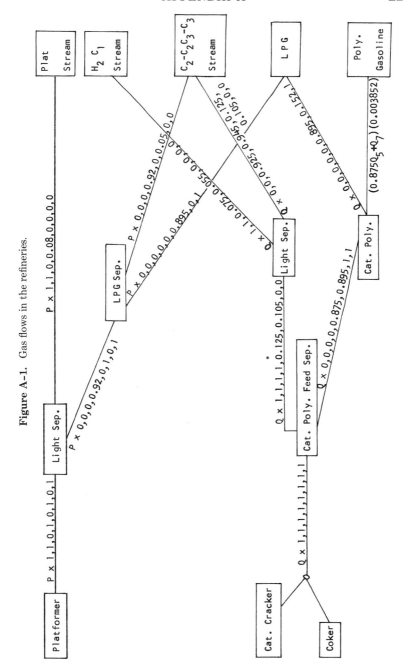

Figure A-1. Gas flows in the refineries.

the stream is referred to as the "C_2—C_2C_3—C_3 stream." (This notation follows customary notation in the literature of petroleum technology and refers to ethylene: C_2—, ethane: C_2, propylene: C_3—, and propane: C_3.)

The cracker–coker united gas stream is denoted in Figure A-1 by $Q \times 1, 1, 1, 1, 1, 1, 1, 1$. This notation corresponds to the notation employed for the platformer stream; the amount of each component in the cracker–coker stream defines the unit level. The stream goes to a separation unit (Cat. Poly. Feed Sep.) which prepares the feed stream for the catalytic-polymerization process. This feed stream contains 87.5% of the propylene, 89.5% of the propane, and all the butylenes and butanes of the cracker–coker stream. The catalytic-polymerization process utilizes only the propylene and butylenes of the feed stream. All the propylene disappears, but part of the butylenes show up in the exit gas stream, which also contains the inert propane and butanes. This exit stream is virtually free of light gases and is assigned directly to LPG output. (Other gases which are produced in insignificant amounts as by-products of the polymerization reaction have been omitted.) The production of polymer gasoline is calculated as a fraction of the combined propylene plus butylenes entering the polymerization unit.

The light stream $Q \times 1, 1, 1, 1, 0.125, 0.105, 0, 0$ from the catalytic-polymerization feed-separation unit is further separated into a stream containing primarily hydrogen and methane (H_2C_1 stream) and into a C_2—C_2C_3—C_3 stream. The assumed separation efficiency is such that 7.5% of the ethylene and 5.5% of the ethane are lost into the H_2C_1 stream.

The sources on which the catalytic-polymerization process inputs and outputs are based are *S1* (p. 134) and *S3* (case tabulations I–VI, pp. 127–132). The reactive inputs to the process consist of propylene and butylene. However, these components have to be diluted with inert gases lest the reaction become too violent. Consequently, a mixture of propylene, propane, butylenes, and butanes constitutes the feed. The relative amounts of these gases are determined by the feed-separation process. Though it would be possible to separate almost all the propylene and propane from the lighter gases, such separation would be uneconomical. It is assumed that an efficient and economical separation plant can separate practically all the butylenes and butanes, 87.5% of the propylene, and 89.5% of the propane (source *S17*). Only the cracker–coker gases are used for catalytic-polymerization feed, since the re-former gases contain no propylene and propane.

In the catalytic-polymerization process, all the propylene disappears, but 15.18% of the butylenes show up in the tail gases. Gasoline output and utilities inputs are based on the weight of feed. Per pound combined propylene plus butylene feed, gasoline = 0.003852 bbl, steam input = 2.02 lb, fuel input = 0, power input = 0.00363 kw-hr. All inputs and outputs except labor are proportional to the amount of combined propylene plus butylene feed. Labor is invariant at 112 man-hours per day between the feed inputs of 280,000 and 550,000 lb/day (Lindsay, *S1*, pp. 113, 118, 134, 139).

Figure A-2 represents the ethylene-separation process as it is formally added to the refinery. As shown in the figure, the input consists of the $C_2-C_2C_3-C_3$ stream. All the incoming ethylene leaves as pure ethylene; all the incoming ethane leaves as pure ethane; all the incoming propane goes to the LPG pool; and all the incoming propylene is fed back to the catalytic-polymerization unit, thus being converted into polymer gasoline (at the rate of 0.003852 bbl per pound propylene).

The utility requirements of the ethylene-separation process just outlined include the extra utility requirement for the increased polymer gasoline production. The utilities are based on (1) pure ethylene produced, $0.925Q_3$ (see Figure A-2), where Q_3 is the ethylene content, in pounds, of the cracker–coker gases, and (2) propylene separated and sent to the catalytic polymerization process, $0.125Q_5$, where Q_5 is the propylene content, in pounds, of the cracker–coker gases. These utilities are:

Steam input: $(8.65)(0.925)Q_3 + (2.02)(0.125)Q_5$
Power input: $(0.0111)(0.925)Q_3 + (0.00363)(0.125)Q_5$

Labor inputs for the ethylene-separation process are included with the labor requirements of activities 14 and 15 (ethylene by cracking). When the latter activities do not occur, the labor input is neglected.

The ethylene-separation process and its relationship to the refinery are highly simplified in defining activities 7 through 12. Furthermore, the separation processes within the refinery are also highly simplified. The sources for all separation-process data were private communications from two different refinery engineering firms. The difficulty in estimating inputs and outputs of separation processes is that the degree of separation is technically variable within wide limits. The separation of the refinery streams in the *absence* of ethylene separation is usually not performed as sharply as indicated above. Moreover, the degree of separation of

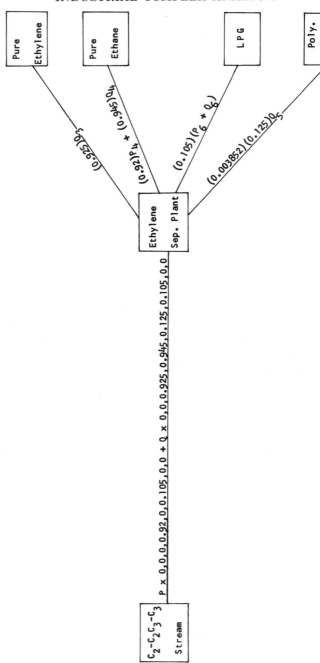

Figure A-2. Formalized gas flows, ethylene separation.

APPENDIX A 227

streams within the refinery depends not only on whether ethylene separation is subsequently performed, but also on whether such commodities as ammonia, acetylene, and hydrogen cyanide are also produced, and, if so, by what process; etc. All of these problems represent refinements the exploration of which was not justified within the scope of the present study.

5. As described in the foregoing sections, refinery and ethylene-separation prototypes are built up in several steps. First, the scheme of basic processes is determined; next, assumptions are made regarding the degree of separation of the various gas streams that would be undertaken; finally, the production of polymer gasoline and the production of ethylene by separation are added on, to the extent permitted by the available feeds.

In Table 3 the data for the activities 1 through 6 are obtained by summing all inputs and outputs of the refinery processes, without, of course, including ethylene separation. The data for activities 7 through 12 are obtained by summing all inputs and outputs of the refinery processes including ethylene separation and by subsequently deducting the inputs and outputs corresponding to the refinery without the ethylene-separation process.

The gas outputs of the refinery activities are shown in Table 3 only by components, not by separate streams. The compositions of the individual streams are used in programming certain petrochemical activities. This procedure is illustrated in Chapter 4, where the streams for refinery 4 are given in detail.

All the numerical values cited in this appendix thus far refer to a *daily* basis. In order to convert to a yearly basis, the daily inputs and outputs are multiplied by 330, which is the assumed number of stream-days (i.e., operating days, allowing for repairs and other shutdowns) per year.

6. The remainder of this appendix deals with two additional activities which are closely interrelated with the processes within the refinery. These activities are the production of benzene and of butadiene.

In Table 3, the production of benzene appears as activity 57. It is noted that the raw material for benzene is straight-run gasoline, while re-formed gasoline appears as a joint product. In fact, the quantity of re-formed gasoline obtained from this activity exceeds the quantity of benzene approximately by a factor of ten.

The process for obtaining benzene consists of re-forming straight-run gasoline by the same process that is used for obtaining premium-grade motor fuel. During this process, various aromatic chemicals, among them benzene, are obtained in mixture with other gasoline components. When the object is the production of motor fuel only, the resulting mixture is used for blending into premium gasoline. When the object is the joint production of aromatic chemicals, the mixture is routed to an extraction step in which the aromatic constituents of the mixture are removed by treatment with a suitable solvent. The remainder is a gasoline of lower quality, since the aromatics are among the highest-grade constituents of premium gasoline. To arrive at the data for activity 57, several assumptions are adopted:

First, it is assumed that benzene is extracted from the suitable portions of the re-formate made from all the available straight-run gasoline and naphtha. Straight-run gasoline from the topping of light crude is 6286 bbl per day. Naphtha from the same source is 2000 bbl per day. The total of 8286 bbl per day is assumed to be re-formed and then sent to aromatics extraction.

Second, it is assumed that only benzene is produced. Thus, any other aromatics obtained as a by-product of the extraction process are assumed to be reunited with the re-formed gasoline. This procedure is feasible, but it may not be desirable, since the joint production of all aromatics—benzene, toluene, xylenes, and heavier elements—may well be economically more desirable than the production of benzene alone. For example, these products may be obtained jointly with benzene and shipped to mainland markets. The problem here is one of differential profitabilities: Is it preferable to produce all aromatics jointly and to ship some of them to mainland markets, incurring a shipping charge that is higher than the one for gasoline; or is it preferable to produce benzene alone, at a somewhat higher unit cost? In the present study, differential profitabilities for refinery products are not examined in detail, and therefore it is assumed that the production of benzene is performed alone.

Third, it is assumed that the production of benzene is undertaken only when a refinery has a re-forming operation for the production of motor fuel. Thus, the total feed to re-forming is 8286 bbl per day; but the benzene activity shows only an input of 3286 bbl per day, since 5000 bbl per day of feed is already included in the refinery activity. Note that only refineries 3, 4, 5, and 6 have a provision for re-forming; thus benzene production by activity 57 is possible only when these refineries are scheduled in a program.

Note further that, when activities 5 and 57 are scheduled simultaneously, the total feed required exceeds the available straight-run gasoline and naphtha, which together amount to only 5000 bbl per day. However, in refinery 5, there is considerable production of coked gasoline which is taken to be a suitable alternative feed and is, in fact, reported in the same commodity row as straight-run gasoline. When coked gasoline is included in the feed, there is still a deficit of $8286 - 5000 - 2854 = 432$ bbl per day of feed. The effect of this deficit is that the maximum scale of activity 57 in conjunction with refinery 5 is $1000 - 432/8286 = 0.948$.

The sources for the inputs and outputs of activity 57 are $S11$ through $S16$ and $S18$ through $S24$.

7. In Table 3, the production of butadiene is represented by two activities: 65 and 66. The sources of data for activity 65 are given in Chapter 3. Activity 65 represents the production of butadiene from refinery butane. Activity 66 is labeled as "butadiene from refinery." This label refers to the fact that certain gaseous components can be diverted from their normal routing and can be used for butadiene production. Such is the case with n-butylene which would normally go to catalytic polymerization in all six refinery prototypes. When n-butylene is routed to butadiene production, the following main-process steps are undertaken: separation of n-butylene from the butane–butylene gases and cracking of n-butylene to butadiene with recirculation of the unreacted portion. In order to obtain the input–output data for activity 66, one must correct the yields and requirements of the above steps for the decrease in catalytic polymerization which takes place when part of the feed to the latter process is used for butadiene production. Consequently, the activity shows an input of polymer gasoline. This input denotes the fact that, when activity 66 is used, butadiene is obtained at the expense of polymer gasoline. The sources for the numerical data for activity 66 consist of the sources given for activity 65 in Chapter 3, and, in addition, the data recorded below on polymer gasoline production.

Both activity 65 and activity 66 operate under restrictions which ensure that the butane or n-butylene used as feed does not exceed the amount actually available. The reason for these special restrictions is that the data on commodity outputs do not distinguish between normal butane and isobutane on the one hand and normal butylene and isobutylene on the other. The special restrictions have been derived by more detailed consideration of the composition of the refinery gases than is given in Table A-2. It is assumed that the ratio of normal butane to isobutane in the refinery gases is 1:2.5. This assumption is based on sources $S25$ and $S26$, which give the ratio as approximately 1:4. However, it is further assumed that about half the total butanes are retained in the gasoline. If no normal butane is used for this purpose, the ratio becomes 1:2.5 as given above. The ratio of normal to isobutylenes is assumed to be 2:1, based on source $S6$. On the basis of these assumptions, the maximum scales of activities 65 and 66 for the six refinery prototypes are calculated from the following data:

Refinery Prototype	1	2	3	4	5	6
Total butanes, M lb/day	37.2	58.8	130.0	151.6	186.2	136.8
Total butylenes, M lb/day	95.1	142.6	118.2	165.8	212.4	142.6
Normal butane, M lb/day	10.6	16.8	37.1	43.3	53.2	39.1
Normal butylenes, M lb/day	63.4	95.1	78.8	110.5	141.6	95.1

Per pound of butadiene, the n-butane input required is 1.82 lb; alternately, per pound of butadiene, the n-butylene input required is 1.37 lb (based on above sources). From these data (converted to a yearly basis by a factor of 330 stream-days/yr) the maximum scales of activities 65 and 66 become, in multiples of unit-level output:

Refinery Prototype	1	2	3	4	5	6
Maximum scale for activity 65	0.19	0.30	0.67	0.79	0.96	0.71
Maximum scale for activity 66	1.53	2.29	1.90	2.66	3.41	2.29

From unit-scale refineries, the maximum annual production of butadiene, in millions of pounds, is:

Refinery Prototype	1	2	3	4	5	6
Maximum butadiene production	17.2	25.9	25.7	34.5	43.7	30.0

8. Sources for the data on refinery and ethylene separation activities are as follows:

S1. John Robert Lindsay, *The Location of Oil Refining in the United States,* unpublished Ph.D dissertation, Harvard University, 1954.

S2. W. L. Nelson, *Petroleum Refinery Engineering,* 3rd ed., McGraw-Hill Book Co., 1949.

S3. T. B. Kimball and J. A. Scott, "Refinery Tools for Producing High Octane Gasoline," *American Petroleum Institute Journal,* **28M** (III), 121-136 (1948).

APPENDIX A

S4. R. T. Carter, "Furfural Refining Scores on Four Counts," *Oil and Gas Journal,* **52,** 161, Table 7 (Mar. 22, 1954).

S5. A. N. Sachanen, *Conversion of Petroleum,* 2nd rev. ed., Reinhold Publishing Corp., 1948, p. 539, Table 198.

S6. E. V. Murphree, "Production of Synthetic Rubbers, Plastics, and Fibers from Petroleum," Association française des techniciens du petrol, Paris, 1949; quoted in R. E. Kirk and D. F. Othmer (eds.), *Encyclopedia of Chemical Technology,* Vol. 10, 1953, "Sources of Hydrocarbon Raw Materials."

S7. Information secured from a large refinery construction firm on inputs and yields of Thermofor Catalytic Cracking.

S8. W. L. Nelson, *Venezuelan and World Crude Oils* (in Spanish and English), Ministerio de Minas e Hidrocarburos, Caracas, 1952.

S9. V. Mekler, A. Schuttle and T. T. Whipple, "Continuous Contact Coking Process Shows Ability to Handle Heavy Feedstocks," *Oil and Gas Journal,* **52,** 200–203 (Nov. 16, 1953), Tables 1 and 3.

S10. Private communication from an engineering firm active in refinery design.

S11. M. J. Fowle, R. D. Bent, and G. P. Masologites, "Tomorrow's Octanes," *American Petroleum Institute Journal,* **32M** (III), 197–207 (1952).

S12. M. J. Fowle, R. D. Bent, B. E. Milner, and G. P. Masologites, "In Reforming, It's the Catalyst That Counts," *Petroleum Refiner,* **31,** 156–169, (Apr. 1952).

S13. J. W. Teter et al., "New Catalytic Reforming Process," *Oil and Gas Journal,* **52,** 118ff (Oct. 12, 1953).

S14. "Fluid Hydroforming," M. W. Kellogg Co., bulletin O-52-3 (1952); and same title, no code number, 1951.

S15. T. A. Burtis and H. D. Knoll, "Houdriforming," *Petroleum Engineer,* **25,** C15–C16 (June 1953).

S16. "An Appraisal of Catalytic Reforming," a special report in *Petroleum Processing,* **10,** 1159–1204 (1955), with extensive bibliography.

S17. Private communication from a large engineering firm engaged in refinery design.

S18. Davis Read, "The Production of High Purity Aromatics for Chemicals," *Petroleum Refiner,* **31,** 97–103 (May 1952).

S19. V. B. Guthrie, "Man-Size Petrochemical Venture," *Petroleum Processing,* **9,** 83 (1954).

S20. V. K. Jackson et al., "Making High Purity Aromatics," *Petroleum Processing,* **9,** 233 (1954).

S21. "Aromatics, Nitration Grade," *Petroleum Processing,* **9,** 1224 (1954).

S22. D. S. Maisel, "Aromatics from Petroleum," *Petroleum Processing,* **8,** 1185–1192 (1953).

S23. "Sun's New Petrochemical Plant," *Chemical Engineering,* **61,** 162 (Feb. 1954).

S24. Utilities for aromatics production: Private communications from petroleum refining firms.

S25. E. C. Oden and J. J. Perry, "What Makes a Good Cat Cracker Feed?" *Oil and Gas Journal,* **53,** 164 (Mar. 22, 1954).

S26. J. W. Schall, J. C. Dart and C. G. Kirkbride, "Moving Bed Catalytic Cracking Correlations," *Chemical Engineering Progress,* **45,** 746–754 (1949).

Appendix B

Specific Dynel, Dacron,
Nylon Subprograms
and Specific
Orlon, Dynel, Dacron,
Nylon Production Programs

APPENDIX B

TABLE B-1
DYNEL SUBPROGRAMS:

DYNEL A

```
Ethylene    11.68⎫                                     
Chlorine    28.54⎭ #23 Ethylene Dichloride 39.58   #29 Vinyl⎫
                                                chloride 23.00⎪
Ethylene    11.76⎫                                           ⎪ #49 Dynel
Chlorine    26.21⎭ #19 Ethylene Oxide 15.64⎫                 ⎬  Polymer
                                           ⎬ #41 Acryloni-   ⎪   36.50
CH₄          9.23⎫                         ⎪    trile 15.33 ⎭     ↓
NH₃          6.65⎭ #39 HCN 7.82           ⎭                    #50 Dynel
                                                                 Staple
                                                                 36.50
```

Total basic inputs: Ethylene 23.44, Chlorine 54.75, CH_4 9.23, NH_3 6.65

Subprograms: (3.958)(#23) + (2.300)(#29) + (3.650)(#49) +
              (1.564)(#19) + (1.533)(#41) + (0.782)(#39) +
              (3.650)(#50)

DYNEL B

```
Ethylene    11.68⎫                                     
Cl₂         28.54⎭ #23 Ethylene dichloride 39.58   #29 Vinyl⎫
                                                chloride 23.00⎪
Ethylene    18.42  #20 Ethylene Oxide 15.64⎫                 ⎪ #49 Dynel
                                           ⎬ #41 Acryloni-   ⎬  Polymer
CH₄          9.23⎫                         ⎪    trile 15.33 ⎭   36.50
NH₃          6.65⎭ #39 HCN 7.82            ⎭                    ↓
                                                                #50 Dynel
                                                                 Staple
                                                                 36.50
```

Total basic inputs: Ethylene 30.10, Chlorine 28.54, CH_4 9.23, NH_3 6.65

Subprogram: (3.958) (#23) + (2.300) (#29) + (3.650) (#49) + (1.564) (#20) +
             (0.782) (#39) + (1.533) (#41).

DYNEL C

```
Ethylene    11.68⎫ #19 Ethylene dichloride 29.58   #29 Vinyl⎫
Cl₂         28.54⎭                              chloride 23.00⎪
                                                              ⎪ #49 Dynel
Acetylene   10.12⎫                                            ⎬  Polymer
CH₄         11.94⎬ #39 HCN 10.12⎫                             ⎪  ↓ 36.50
NH₃          8.60⎭              ⎬ #40 Acryloni-              ⎭ #50 Dynel
                                     trile 15.33                 Staple
                                                                 36.50
```

Total basic inputs: Ethylene 11.68, Acetylene 10.12, Cl_2 28.54, CH_4 11.94, NH_3 8.60

Subprogram: (3.958) (#23) + (2.300) (#29) + (3.650) (#49) + (3.650) (#50) +
             (1.012) (#39) + (1.533) (#40).

TABLE B-1 (Cont'd)
DYNEL SUBPROGRAMS:

DYNEL D

Acetylene	9.89 ⎫	#30		
HCl	15.87 ⎭	Vinyl Chloride 23.00		#49 Dynel Polymer
Ethylene	11.76 ⎫	#19		36.50
Chlorine	26.21 ⎭	Ethylene Oxide 15.64 ⎫	#41 Acrylon-	#50 ↓
CH_4	9.23 ⎫	#39	itrile 15.33 ⎭	Dynel Staple
NH_3	6.65 ⎭	HCN 7.82		36.50

Basic inputs: Acetylene 9.89, Ethylene 11.76, HCl 15.87, Chlorine 26.21, CH_4 9.23, NH_3 6.65.

Subprogram: (2.300) (#30) + (3.650) (#49) + (3.650) (#50) + (1.564) (#19) + (0.782) (#39) + (1.533) (#41).

DYNEL E

Acetylene	9.89 ⎫	#30		
HCl	15.87 ⎭	Vinyl Chloride 23.00		#49 Dynel Polymer
Ethylene	18.42	#20 Ethylene Oxide 15.64 ⎫	#41 Acrylon-	36.50
CH_4	9.23 ⎫	#39	itrile 15.33 ⎭	#50 ↓ Dynel
NH_3	6.65 ⎭	HCN 7.82		Staple 36.50

Basic inputs: Acetylene 9.89, HCl 15.87, Ethylene 18.42, CH_4 9.23, NH_3 6.65.

Subprogram: (2.300) (#30) + (3.650) (#49) + (3.650) (#50) + (1.564) (#20) + (0.782) (#39) + (1.533) (#41).

DYNEL F

Acetylene	9.89 ⎫	#30		
HCl	15.87 ⎭	Vinyl chloride 23.00		#49 Dynel Polymer
Acetylene	10.12		#40 Acrylonitrile	↓ 36.50
CH_4	11.94 ⎫	#39	15.33	#50 Dynel Staple
NH_3	8.60 ⎭	HCN 10.12		36.50

Basic inputs: Acetylene 20.01, HCl 15.87, CH_4 11.94, NH_3 8.60

Subprogram: (2.300) (#30) + (3.650) (#49) + (3.650) (#50) + (1.012) (#39) + (1.533) (#40).

APPENDIX B

TABLE B-2
DACRON SUBPROGRAMS

DACRON A

Ethylene	9.79	#22 Et. Glycol	11.79⎱	#46 Dacron	#47 Dacron
P-xylene	25.07	#44 D. M. T.	36.87⎰	Polymer 36.5	staple 36.5

Subprogram: $(3.65)(\#47) + (3.65)(\#46) + (1.179)(\#22) + (3.687)(\#44)$

DACRON B

Ethylene	6.98⎫				
Cl$_2$	15.55⎭	#21 Et. Glycol	11.79⎱	#46 Dacron	#47 Dacron
P-xylene	25.07	#44 D. M. T.	36.87⎰	Polymer 36.5	staple 36.5

Subprogram: $(3.65)(\#47) + (3.65)(\#46) + (1.179)(\#21) + (3.687)(\#44)$

DACRON C

Ethylene	9.79⎫				
P-xylene	21.75⎭	#22 Et. Glycol	11.79⎱	#46 Dacron	#47 Dacron
HNO$_3$	43.51⎫	#45 D. M. T.	36.87⎰	Polymer 36.5	staple 36.5
H$_2$SO$_4$	5.61⎭				

Subprogram: $(3.65)(\#47) + (3.65)(\#46) + (1.179)(\#22) + (3.687)(\#45)$

DACRON D

Ethylene	6.89⎫				
Cl$_2$	15.55⎭	#21 Et. Glycol	11.79⎱		
P-xylene	21.75⎫			#46 Dacron	#47 Dacron
HNO$_3$	43.51⎬	#45 D. M. T.	36.87⎰	Polymer 36.5	staple 36.5
H$_2$SO$_4$	5.16⎭				

Subprogram: $(3.65)(\#47) + (3.65)(\#46) + (1.179)(\#21) + (3.687)(\#45)$

TABLE B-3
NYLON SUBPROGRAMS

NYLON A-F ("Route A")

Adiponitrile   19.15 ⎫   #70 Hexamethylene ⎫
H_2 equivalent ammonia   6.85 ⎭   diamine   18.72 ⎬ #71 Nylon
                                        Adipic Acid 23.58 ⎭   │ Salt
                                                               ↓ 36.5
                                                               #72 Nylon
                                                                Staple
                                                                36.5

Subprogram: (3.650) (#72) + (3.650) (#71) + (1.872) (#70)

NYLON G-H ("Route C")

Adipic Acid 34.87   #68 Adiponitrile   19.15 ⎫   #70 Hexamethylene ⎫ #71 Nylon
                     H_2 equivalent ammonia 6.85 ⎭   diamine 18.72 ⎬ Salt
                                                          Adipic Acid 23.58 ⎭   36.5
                                                                                #72 Nylon
                                                                                 Staple
                                                                                 36.5

Subprogram: (3.650)(#72) + (3.650)(#71) + (1.872)(#70) + (1.915)(#68)

NYLON J-K ("Route B")

Adiponitrile   19.15 ⎫   #70 Hexamethylene ⎫
H_2 equivalent ammonia   6.85 ⎭   diamine   18.72 ⎬ #71 Nylon
Adiponitrile   19.36   #69 Adipic Acid 23.58 ⎭   │ Salt
                                                               ↓ 36.5
                                                               #72 Nylon
                                                                Staple
                                                                36.5

Subprogram: (3.650)(#72) + (3.650)(#71) + (1.872)(#70) + (2.358)(#69).

APPENDIX B

TABLE B-4
PRODUCTION PROGRAM: ORLON B

	Subprogram	
		(3.650) (#52)
		(3.650) (#53)
(PRODUCTION PROGRAMS FOR ORLON A, C, E,		(3.833) (#41)
H, ARE GIVEN AFTER ORLON B, D, F, J.)		(3.911) (#20)
		(1.955) (#39)

<u>Ethylene</u> need = 46.06. Based on refinery #4.   (1.000) (#4)
By sep. (#10) available 16.10. Ethane by-product 30.19   (1.000) (#10)
By ethane cracking (#14) 23.40. Ethane consumed 30.19   (2.340) (#14)
By propane cracking (#15) <u>6.56.</u> Propane consumed 19.55   (0.656) (#15)
 Total ethylene produced 46.06

<u>Propane</u> need for cracking to ethylene = 19.55. Program
 has to include dummy activity #13: propane from   (1.955) (#13)
 L. P. G.

<u>Methane</u> need for subprogram = 23.08. In H_2C_1 stream,
 available 31.11; use 23.08/31.11 = 0.742 of H_2C_1
 stream. Leftover = 0.258.

<u>Ammonia</u> need for subprogram = 16.63
           for fertilizer = <u>80.00</u>
                            96.63

   Production from PLAT −43.91
        from leftover H_2C_1 <u>−18.11</u>
   balance from cycle oil   34.61

NH_3 from:	PLAT	0.258H_2C_1	Cycle Oil	TOTAL		
hydrogen	35.95	2.31	0	38.26	(3.826)	(#31)
methane	6.82	14.59	0	21.41	(2.141)	(#32)
ethylene	0	0.58	0	0.58	(0.058)	(#33)
ethane	1.14	0.63	0	1.77	(0.177)	(#34)
cycle oil	0	0	<u>34.61</u>	<u>34.61</u>	(3.461)	(#38)
	43.91	18.11	34.61	96.63		

Fertilizer: standard   (8.760) (#55)
                       (6.900) (#56)
                       (6.690) (#43)

TABLE B-5
PRODUCTION PROGRAM: ORLON D

	Subprogram	
	(3.650)	(#52)
	(3.650)	(#53)
	(3.833)	(#41)
	(3.911)	(#20)
	(1.955)	(#39)

<u>Ethylene</u> need = 46.06. Based on #4 refinery at a scale sufficient to eliminate need for propane cracking. (1.166) (#4)
By separation (1.166)(#10) available 18.77 (1.166) (#10)
 Ethane by-product 35.20
By ethane cracking (#14)   <u>27.29</u> (2.729) (#14)
 Ethane consumed 35.20
Total ethylene produced   46.06

<u>Methane</u> need = 23.08. In H_2C_1 stream, available (31.11) (1.166). Use 23.08/(31.11 × 1.166) of H_2C_1 stream of the scale = 1.166 refinery. This is the same thing as 23.08/31.11 = 0.742 of the unit H_2C_1 stream of the unit scale refinery. Leftover = 1.166 − 0.742 = 0.424 of the unit H_2C_1 stream.

<u>Ammonia</u> need for subprogram 16.63
 for fertilizer <u>80.00</u>
   96.63

Production from 1.166 x
   unit PLAT −51.20
from 0.424 of unit H_2C_1 <u>−29.75</u>
 balance from cycle oil: 15.69

NH_3 from:	PLAT	0.424H_2C_1	Cycle Oil	TOTAL		
hydrogen	41.92	3.78	0	45.70	(4.570)	(#31)
methane	7.95	23.98	0	31.93	(3.193)	(#32)
ethylene	0	0.96	0	0.96	(0.096)	(#33)
ethane	1.33	1.03	0	2.36	(0.236)	(#34)
cycle oil	0	0	15.69	15.69	(1.569)	(#38)
	51.20	29.75	15.69	96.64		

<u>Fertilizer</u>: standard

(8.760) (#55)
(6.900) (#56)
(6.690) (#43)

APPENDIX B

TABLE B-6
PRODUCTION PROGRAM: ORLON F

	Subprogram	
	(3.650)	(#52)
	(3.650)	(#53)
	(3.833)	(#41)
	(3.911)	(#19)
	(1.955)	(#39)

<u>Chlorine</u> need = 65.53. Made by electrolysis.
By-product H_2 = 1.90 (6.553) (#16)

<u>Lime</u> need for #19 = 52.33. Add lime production. (5.233) (#18)

<u>Ethylene</u> need = 29.40. Based on #4 refinery. (1.000) (#4)
By sep. (#10) available 16.10. By-product ethane 30.19 (1.000) (#10)
By ethane cracking (#14) <u>13.30</u>. Ethane consumed <u>17.16</u> (1.330) (#14)
Total ethylene produced 29.40. Leftover ethane 13.03

<u>Methane</u> need = 23.08. In H_2C_1 stream, available 31.11;
use 23.08/31.11 = 0.742 of H_2C_1 stream. Leftover = 0.258.

<u>Ammonia</u> need for subprogram 16.63
        for fertilizer <u>80.00</u>
                           96.63
Produced from by-product
        hydrogen   0   (see notes)
    from PLAT   -43.91
    from 0.258 H_2C_1 -18.11
    from leftover
        ethane   <u>-22.54</u>
balance from cycle oil 12.07

NH_3 from:	BPH_2	PLAT	$0.258H_2C_1$	leftover ethane	cycle oil	TOTAL		
hydrogen	0	35.95	2.31	0	0	38.26	(3.826)	(#31)
methane	0	6.82	14.59	0	0	21.41	(2.141)	(#32)
ethylene	0	0	0.58	0	0	0.58	(0.058)	(#33)
ethane	0	1.14	0.63	22.54	0	24.31	(2.431)	(#34)
cycle oil	0	0	0	0	12.07	12.07	(1.207)	(#38)
		43.91	18.11	22.54	12.07	96.63		

<u>Fertilizer</u>: standard

	(8.760)	(#55)
	(6.900)	(#56)
	(6.690)	(#43)

<u>Notes</u>: By-product hydrogen should have been programmed for ammonia, thus reducing ammonia production from cycle oil. (Error is slight.)

TABLE B-7
PRODUCTION PROGRAM: ORLON J

	Subprogram		
		(3.650)	(#52)
		(3.650)	(#53)
		(3.833)	(#40)
		(2.530)	(#39)

Methane need = 29.85. Based on refinery #4. In H_2C_1 stream, available 31.11; use 29.85/31.11 = 0.960 of H_2C_1 stream. Leftover = 0.040

(1.000) (#4)

Acetylene need = 23.50. Made by cracking $C_2-C_2C_3-C_3$ stream. Unit stream yields 29.60 acetylene; so use 25.30/29.60 = 0.856 of unit stream. Leftover = 0.144 of stream.

Acetylene from: 0.856 $C_2-C_2C_3-C_3$

ethylene	7.62	(0.762)	(#25)
ethane	13.17	(1.317)	(#26)
propylene	2.75	(0.275)	(#27)
propane	1.76	(0.176)	(#28)
	25.30		

Ammonia need for subprogram 21.50
         for fertilizer <u>80.00</u>
                              101.50
Production from PLAT -43.91
      from 0.040 H_2C_1 - 2.80
      from 0.144 $C_2-C_2C_3-C_3$ <u>-14.12</u>
balance from cycle oil 40.67

NH_3 from:	PLAT	0.040H_2C_1	0.144C_2-C_2C_3-C_3	cycle oil	TOTAL		
hydrogen	35.95	0.36	0	0	36.31	(3.631)	(#31)
methane	6.82	2.26	0	0	9.08	(0.908)	(#32)
ethylene	0	0.08	3.69	0	3.77	(0.377)	(#33)
ethane	1.14	0.10	7.52	0	8.76	(0.876)	(#34)
propylene	0	0	1.70	0	1.70	(0.170)	(#35)
propane	0	0	1.21	0	1.21	(0.121)	(#36)
cycle oil	0	0	0	40.67	40.67	(4.067)	(#38)
	43.91	2.80	14.12	40.67	101.50		

Fertilizer: standard

(8.760) (#55)
(6.900) (#56)
(6.690) (#43)

APPENDIX B

TABLE B-8
PRODUCTION PROGRAMS: ORLON A, C, E, H

These programs differ from the programs for ORLON B, D, F, and J in that the ammonia production activities from cycle oil were not included. This reduces the total ammonia available for fertilizer and hence the scale of the fertilizer activities. The latter scales are proportional to the total ammonia available for fertilizer in each case.

The tabulation below shows the standard case where ammonia for fertilizer = 80, and the changes for ORLON A, C, E, and H.

	Standard	ORLON			
		A	C	E	H
Total ammonia available for fertilizer	80.00	45.39	64.31	67.93	39.33
Scale of activity #55	8.760	4.975	7.048	7.445	4.310
#56	6.900	3.913	5.544	5.856	3.395
#43	6.690	3.799	5.383	5.686	3.291

INSTRUCTIONS for changing ORLON B to ORLON A
ORLON D to ORLON C
ORLON F to ORLON E
ORLON J to ORLON H

(1) Omit from the programs of ORLON B, D, F, or J, respectively, the activity scales for #38, #55, #56, and #43. (These are the last four items on the production program sheets.)

(2) Add activities #55, #56, and #43 at the scales shown in the above tabulation.

NOTE: Ammonia activity (including need for subprogram) slightly low for ORLON A, H.

TABLE B-9
PRODUCTION PROGRAM: DYNEL A

Subprogram	(3.958)	(#23)
	(2.300)	(#29)
	(3.650)	(#49)
	(1.564)	(#19)
	(1.533)	(#41)
	(0.782)	(#39)
	(3.650)	(#50)

Chlorine need = 54.75. Add chlorine electrolysis. (5.475) (#16)
  By-product hydrogen = 1.56
For activity #19, lime need = 20.93. Add lime pro- (2.093) (#18)
  duction.

Ethylene need = 23.44. Based on refinery #4   (1.000) (#4)
By sep. (#10) available   16.10. Ethane by-product 30.19   (1.000) (#10)
By ethane cracking (#14) +7.34. Ethane consumed   -9.46   (0.734) (#14)
Total ethylene produced   23.44   Ethane leftover   20.73

Methane need = 9.23. In H_2C_1 stream, available 31.11.

  Use 9.23/31.11 = 0.297 of H_2C_1 stream. Leftover
  H_2C_1 = 0.703

Ammonia   for subprogram   6.65
                  for fertilizer 80.00
                                  86.65
Production from by-product
                       H_2 0   (see notes) (3)
           from PLAT   -43.91
  balance from H_2C_1 42.74

Unit H_2C_1 stream yields 70.01 NH_3; so use 42.74/70.01 =
  0.601 H_2C_1

NH_3 from:	H_2	PLAT	0.601 H_2C_1	TOTAL		
hydrogen	0	35.95	5.46	41.41	(4.141)	(#31)
methane	0	6.82	34.53	41.35	(4.135)	(#32)
ethylene	0	0	1.27	1.27	(0.127)	(#33)
ethane	0	1.14	1.48	2.62	(0.262)	(#34)
		43.91	42.74	86.65		

Fertilizer: standard   (8.760) (#55)
                                                   (6.900) (#56)
                                                   (6.690) (#43)

Notes: (1) Ethylene plant slightly low.
        (2) HCN plant (#39) low.
        (3) By-product should have been programmed
            for ammonia, thus reducing H_2C_1 going to
            ammonia. (Error is slight.)

APPENDIX B

TABLE B-10
PRODUCTION PROGRAM: DYNEL B

	Subprogram		
		(3.958)	(#23)
		(2.300)	(#29)
		(3.650)	(#49)
		(3.650)	(#50)
		(1.564)	(#20)
		(0.782)	(#39)
		(1.533)	(#41)

Chlorine need = 28.54. Add chlorine electrolysis. (2.854) (#16)
By-product hydrogen = 0.81.

Ethylene need = 30.10. Based on refinery #4 (1.000) (#4)
By sep. (#10) available 16.10. Ethane by-product 30.19 (1.000) (#10)
By ethane cracking (#14)+14.00. Ethane consumed -18.05 (1.400) (#14)
Total ethylene produced 30.10. Ethane leftover 12.14

Methane need = 9.23. In H_2C_1 stream available 31.11.

Use 9.23/31.11 = 0.297 of H_2C_1 stream. Leftover
H_2C_1 = 0.703.

Ammonia need for subprogram 6.65
        for fertilizer 80.00
                              86.65

Production from by-product
                 H_2 0   (see notes)
    from PLAT   -43.91
balance from H_2C_1 42.74

Unit H_2C_1 stream yields 70.01 NH_3; so use 42.74/70.01 = 0.601 H_2C_1

NH_3 from:	H_2	PLAT	0.601 H_2C_1	TOTAL		
hydrogen	0	35.95	5.46	41.41	(4.141)	(#31)
methane	0	6.82	34.53	41.35	(4.135)	(#32)
ethylene	0	0	1.27	1.27	(0.127)	(#33)
ethane	0	1.14	1.48	2.62	(0.262)	(#34)
		43.91	42.74	86.65		

Fertilizer: standard (8.760) (#55)
                              (6.900) (#56)
                              (6.690) (#43)

Notes: (1) HCN plant (#39) low.
        (2) By-product H_2 should have been programmed for ammonia, thus reducing H_2C_1 going to ammonia. (Error is slight.)

TABLE B-11
PRODUCTION PROGRAM: DYNEL C

	Subprogram	
	(3.958)	(#23)
	(2.300)	(#29)
	(3.650)	(#49)
	(3.650)	(#50)
	(1.012)	(#39)
	(1.533)	(#40)

Chlorine need = 28.54. Add chlorine electrolysis. By-product hydrogen = 0.81.　　　　　　　　　　　(2.854) (#16)

Ethylene need = 11.68. Based on refinery #4.
Ethylene separation (#10) at unit scale gives　　　　(1.000) (#4)
16.10 ethylene, therefore, pursue at scale 11.68/
16.10 = 0.725. Leftover $C_2-C_2C_3-C_3$ stream =　　(0.725) (#10)
0.275. By-product ethane = 21.89.

Methane need = 11.94. In H_2C_1 stream, available 31.11.
Use 11.94/31.11 = 0.385 of H_2C_1 stream. Leftover
H_2C_1 = 0.615.

Acetylene need = 10.12. Produced by cracking ethane
by-product of ethylene separation. Ethane consumed = 19.75. Leftover ethane = 21.89 - 19.75　　　(1.012) (#26)
= 2.14.

Ammonia need for subprogram　8.60
　　　　　　 for fertilizer　80.00
　　　　　　　　　　　　　　　88.60
Production from by-product
　　　　　　　　　　H_2　　0　　(see notes)
　　from PLAT　　　　　　-43.91
from leftover 0.615 H_2C_1　-43.05
　　　　　　　　　　　　　　1.64.　Uses 1.81 ethane <
balance from leftover by-　　　　2.14.
　　product ethane

NH_3 from:	H_2	PLAT	0.615H_2C_1	1.81 B.P. ethane	TOTAL		
hydrogen	0	35.95	5.50		41.45	(4.145)	(#31)
methane	0	6.82	34.78		41.60	(4.160)	(#32)
ethylene	0	0	1.28		1.28	(0.128)	(#33)
ethane	0	1.14	1.49	1.64	4.27	(0.427)	(#34)
		43.91	43.05	1.64	88.60		

Fertilizer: standard　　　　　　　　　　　　　　　　(8.760) (#55)
　　　　　　　　　　　　　　　　　　　　　　　　　　　(6.900) (#56)
　　　　　　　　　　　　　　　　　　　　　　　　　　　(6.690) (#43)

Notes: (1) Ethylene plant very low.
　　　　(2) By-product H_2 should have been programmed
　　　　　　for ammonia, thus reducing ethane going to
　　　　　　ammonia. (Error is slight.)

APPENDIX B

TABLE B-12
PRODUCTION PROGRAM: DYNEL D

	Subprogram	
	(2.300)	(#30)
	(3.650)	(#49)
	(3.650)	(#50)
	(1.564)	(#19)
	(0.782)	(#39)
	(1.533)	(#41)

<u>Hydrochloric acid</u> need = 15.87. Based on chlorine and hydrogen, both derived in needed amount by electrolysis.   (1.587) (#17)

<u>Chlorine</u> need = 26.21 for subprogram plus 15.47 for HCL, total 41.68. Made by electrolysis. By-product hydrogen available for outside use = 0.75.   (4.168) (#16)

<u>Lime</u> need = 20.93, for activity #19. Add lime production   (2.093) (#18)

<u>Ethylene</u> need = 11.76. Based on refinery #4   (1.000) (#4)
Ethylene separation at unit scale gives 16.10 ethylene, therefore pursue at scale 11.76/16.10 = 0.730. Leftover C_2-C_2C_3-C_3 stream = 0.270. By-product ethane = 22.04   (0.730) (#10)

<u>Methane</u> need = 9.23. In H_2C_1 stream, available 31.11. Use 9.23/31.11 = 0.297 of H_2C_1 stream. Leftover H_2C_1 = 0.703.

<u>Acetylene</u> need = 9.89. Based on cracking 19.28 ethane by-product of ethylene separation. Leftover ethane = 2.76.   (0.989) (#26)

<u>Ammonia</u> need for subprogram   6.65
       for fertilizer <u>80.00</u>
                         86.65

Produced from by-product
      hydrogen   0   (see notes)
      from PLAT   <u>-43.91</u>
balance from leftover H_2C_1 42.74

Unit H_2C_1 stream yields 70.01 NH_3; so use 42.74/70.01 = 0.601 H_2C_1.

NH_3 from:	H_2	PLAT	0.601 H_2C_1	TOTAL		
hydrogen	0	35.95	5.46	41.41	(4.141)	(#31)
methane	0	6.82	34.53	41.35	(4.135)	(#32)
ethylene	0	0	1.27	1.27	(0.127)	(#33)
ethane	0	<u>1.14</u>	<u>1.48</u>	<u>2.62</u>	(0.262)	(#34)
		43.91	42.74	86.65		

<u>Fertilizer</u>: standard   (8.760) (#55)
                                 (6.900) (#56)
                                 (6.690) (#43)

Notes: (1) Ethylene plant very low.
       (2) Acetylene slightly low.
       (3) By-product H_2 should have been programmed for ammonia, thus reducing H_2C_1 going to ammonia. (The error is slight.)

TABLE B-13
PRODUCTION PROGRAM: DYNEL E

	Subprogram		
		(2.300)	(#30)
		(3.650)	(#49)
		(3.650)	(#50)
		(1.564)	(#20)
		(0.782)	(#39)
		(1.533)	(#41)

Hydrochloric acid need = 15.87. Based on chlorine and hydrogen, both derived in needed amounts by electrolysis.   (1.587) (#17)

Chlorine need = 15.47 for HCL. By-product H_2 used for HCL.   (1.547) (#16)

Ethylene need = 18.42. Based on refinery #4. Ethylene separation at unit scale gives 16.10 ethylene. When refinery is increased in ratio 18.42/16.10 = 1.144, ethylene by separation (#10) at this same scale covers need and no cracking is required. By-products ethane = 34.54.   (1.144) (#4)
   (1.144) (#10)

Methane need = 9.23. In H_2C_1 stream, available (31.11) (1.144). Use 9.23/(31.11 × 1.144) of H_2C_1 stream of the scale = 1.144 refinery. This is the same thing as 9.23/31.11 = 0.297 of the unit H_2C_1 stream of the unit scale refinery. Leftover = 1.144 - 0.297 = 0.847 of the unit H_2C_1 stream.

Acetylene need = 9.89; by cracking 19.29 ethane. Leftover ethane = 34.54 - 19.29 = 15.25.   (0.989) (#26)

Ammonia needed for subprogram   6.65
               for fertilizer   80.00
                                    ─────
                                    86.65
Production from PLAT -(1.144)(43.91) -50.23
                                    ─────
balance from H_2C_1 36.42

Unit H_2C_1 stream yields 70.01 NH_3; so use 36.42/70.01 = 0.520 of unit H_2C_1 stream.

NH_3 from:	1.140 of unit PLAT	0.520 of unit H_2C_1	TOTAL		
hydrogen	41.13	4.66	45.79	(4.579)	(#31)
methane	7.80	29.42	37.22	(3.722)	(#32)
ethylene	0	1.08	1.08	(0.108)	(#33)
ethane	1.30	1.26	2.56	(0.256)	(#34)
	50.23	36.42	86.65		

Fertilizer: standard   (8.760) (#55)
   (6.900) (#56)
   (6.690) (#43)

Notes: (1) Ethylene plant low.
       (2) HCN (#39) low.
       (3) Acetylene (#26) slightly low.

APPENDIX B

TABLE B-14
PRODUCTION PROGRAM: DYNEL F

	Subprogram		
		(2.300)	(#30)
		(3.650)	(#49)
		(3.650)	(#50)
		(1.012)	(#39)
		(1.533)	(#40)

<u>Hydrochloric acid</u> need = 15.87. Based on chlorine and hydrogen, both derived in needed amounts by electrolysis.   (1.587) (#17)

<u>Chlorine</u> need = 15.47 for HCL. Hydrogen by-product used for HCL.   (1.547) (#16)

<u>Methane</u> need = 11.94. Based on unit refinery #4. In H_2C_1 stream, there is 31.11 methane. Use 11.94/31.11 = 0.385 of H_2C_1. Leftover 0.615.   (1.000) (#4)

<u>Acetylene</u> need = 20.01. Made by cracking $C_2-C_2C_3-C_3$ stream. Unit stream yields 29.60 acetylene, so use 20.01/29.60 = 0.676 of unit $C_2-C_2C_3-C_3$ stream. Leftover 0.324 of stream.

Acetylene from:	0.676 $C_2-C_2C_3-C_3$		
ethylene	5.981	(0.598)	(#25)
ethane	10.466	(1.047)	(#26)
propylene	2.172	(0.217)	(#27)
propane	1.391	(0.139)	(#28)
	20.010		

Ammonia need for subprogram	8.60
for fertilizer	80.00
	88.60
Production from PLAT	-43.91
from leftover H_2C_1	-43.05
balance from leftover $C_2-C_2C_3-C_3$	1.64

Unit $C_2-C_2C_3-C_3$ stream yields 98.04 NH_3; so use 1.64/98.04 = 0.017 of unit stream.

NH_3 from:	PLAT	leftover H_2C_1	0.017 $C_2-C_2C_3-C_3$	TOTAL		
hydrogen	35.95	5.50	0	41.45	(4.145)	(#31)
methane	6.82	34.78	0	41.60	(4.160)	(#32)
ethylene	0	1.28	0.43	1.71	(0.171)	(#33)
ethane	1.14	1.49	0.88	3.51	(0.351)	(#34)
propylene	0	0	0.20	0.20	(0.020)	(#35)
propane	0	0	0.14	0.14	(0.014)	(#36)
	43.91	43.05	1.64	88.60		

<u>Fertilizer</u>: standard   (8.760) (#55)
    (6.900) (#56)
    (6.690) (#43)

TABLE B-15
PRODUCTION PROGRAM: DACRON A

			Subprogram		
				(3.650)	(#47)
				(3.650)	(#46)
				(1.179)	(#22)
				(3.687)	(#44)

<u>Ethylene</u> need = 9.79. Based on refinery #4. By separation at unit scale available 16.10, so pursue at scale 9.79/16.10 = 0.608. By-product ethane = 18.36

                                                                          (1.000) (#4)
                                                                          (0.608) (#10)

<u>Ammonia</u> need for fertilizer 80.00
    Produced from PLAT   -43.91
    balance from H_2C_1 36.06

Unit H_2C_1 stream yields 70.01 NH_3; so use 36.06/70.01 = 0.515 of unit stream.

NH_3 from:	PLAT	0.515H_2C_1	TOTAL		
hydrogen	35.95	4.61	40.56	(4.056)	(#31)
methane	6.82	29.13	35.95	(3.595)	(#32)
ethylene	0	1.07	1.07	(0.107)	(#33)
ethane	1.14	1.25	2.39	(0.239)	(#34)
	43.91	36.06	79.97		

<u>Fertilizer</u>: standard

                                                                          (8.760) (#55)
                                                                          (6.900) (#56)
                                                                          (6.690) (#43)

<u>Note</u>: Ethylene plant very low.

APPENDIX B

TABLE B-16
PRODUCTION PROGRAM: DACRON B

	Subprogram		
		(3.650)	(#47)
		(3.650)	(#46)
		(1.179)	(#21)
		(3.687)	(#44)

Chlorine need = 15.55. Made by electrolysis. Hydrogen by-product = 0.451. (1.555) (#16)

Lime need for #21 = 12.41. Add lime production (1.241) (#18)

Ethylene need = 6.98. Based on refinery #4. Ethylene separation (#10) at unit scale give 16.10, so pursue at scale 6.98/16.10 = 0.434. By-product ethane = 13.10. (1.000) (#4)
(0.434) (#10)

```
Ammonia need for fertilizer          80.00
   Produced from by-product hy-
        drogen                       -2.26
        from PLAT                   -43.91
balance from H₂C₁                    33.81
```

Unit H_2C_1 stream yields 70.01 NH_3, so use 33.81/70.01 = 0.483 of unit stream.

NH_3 from:	Hydrogen	PLAT	0.483H_2C_1	TOTAL		
hydrogen	2.26	35.95	4.32	42.53	(4.253)	(#31)
methane	0	6.82	27.32	34.14	(3.414)	(#32)
ethylene	0	0	1.00	1.00	(0.100)	(#33)
ethane	0	1.14	1.17	2.31	(0.231)	(#34)
	2.26	43.91	33.81	79.98		

Fertilizer: standard (8.760) (#55)
(6.900) (#56)
(6.690) (#43)

Notes: (1) Ethylene plant very low.

TABLE B-17
PRODUCTION PROGRAM: DACRON C

	Subprogram		
		(3.650)	(#47)
		(3.650)	(#46)
		(1.179)	(#22)
		(3.687)	(#45)
Nitric acid need for subprogram 43.51			
for fertilizer 66.90			
110.41		(11.041)	(#43)
Nitric acid based on ammonia. Extra ammonia need (over and above fertilizer) = 12.44			
Sulfuric Acid need = 5.16. Based on sulfur 1.78		(0.516)	(#42)
Ethylene need = 9.79. Based on refinery #4. Ethylene separation at unit level yields 16.10, so pursue scale 9.79/16.10 = 0.608.		(1.000)	(#4)
		(0.608)	(#10)

Ammonia need for nitric acid 12.44
     for fertilizer 80.00
             92.44
  Production from PLAT −43.91
             48.52

Unit H_2C_1 stream yields 70.01 NH_3, so use 48.52/70.01 = 0.693 of unit stream.

NH_3 from:	PLAT	$0.693H_2C_1$	TOTAL		
hydrogen	35.95	6.20	42.15	(4.215)	(#31)
methane	6.82	39.20	46.02	(4.602)	(#32)
ethylene	0	1.44	1.44	(0.144)	(#33)
ethane	1.14	1.68	2.82	(0.282)	(#34)
	43.91	48.52	92.43		

Fertilizer: standard, except nitric acid requirement already stated above.	(8.760)	(#55)
	(6.900)	(#56)

Notes: (1) Sulfuric acid (#42) very low.
     (2) Ethylene plant very low.

APPENDIX B 251

TABLE B-18
PRODUCTION PROGRAM: DACRON D

	Subprogram
	(3.650) (#47)
	(3.650) (#46)
	(1.179) (#21)
	(3.687) (#45)

<u>Chlorine</u> need = 15.55. Made by electrolysis. Hydrogen by-product = 0.451.   (1.555) (#16)

<u>Lime</u> need for #21 = 12.41. Add lime production.   (1.241) (#18)

<u>Nitric acid</u> need for subprogram 43.51
for fertilizer 66.90
110.41
Nitric acid based on ammonia. Extra ammonia need (over and above fertilizer) = 12.44.   (11.041) (#43)

<u>Sulfuric acid</u> need = 5.16. Based on sulfur 1.78.   (0.516) (#42)

<u>Ethylene</u> need = 6.98. Based on refinery #4. Ethylene separation (#10) at unit level yields 16.10, so pursue at scale 6.98/16.10 = 0.434. Ethane by-product = 13.10.   (1.000) (#4)
  (0.434) (#10)

<u>Ammonia</u> need for nitric acid 12.44
for fertilizer 80.00
92.44
Production from by-product H_2 -2.26
from PLAT -43.91
balance from H_2C_1 46.28

Unit H_2C_1 stream yields 70.01 NH_3, so use 46.28/70.01 = 0.661 H_2C_1.

NH_3 from:	B.P. Hydrogen	PLAT	0.661 H_2C_1	TOTAL	
hydrogen	2.26	35.95	5.92	44.13	(4.413) (#31)
methane	0	6.82	37.39	44.21	(4.421) (#32)
ethylene	0	0	1.37	1.37	(0.137) (#33)
ethane	0	1.14	1.60	2.74	(0.274) (#34)
	2.26	43.91	46.28	92.45	

<u>Fertilizer</u>: standard, except nitric acid requirement already stated above.   (8.760) (#55)
  (6.900) (#56)

<u>Notes</u>: (1) Sulfuric acid (#42) very low.
(2) Ethylene plant very low.

TABLE B-19
PRODUCTION PROGRAM: NYLON A
(mixed base)

	Staple	(3.650)	(#72)
Subprogram	Salt	(3.650)	(#71)
	HMD	(1.872)	(#70)

Adiponitrile from tetrahydrofuran   (1.915) (#64)
Tetrahydrofuran from furfural   (1.679) (#62)

Furfural	28.43MM	not scheduled, since fur-
Bagasse	625MM	fural differential taken
Sulfuric acid	6.40MM	directly

Sodium cyanide   18.46MM from HCN   (1.846) (#63)
Hydrogen cyanide   10.15MM from methane   11.98   (1.015) (#39)
                             and ammonia   8.63

Sodium hydroxide for sodium cyanide   15.14MM

Hydrochloric acid for adiponitrile, from chlorine   (1.703) (#17)
Chlorine by electrolysis   (1.661) (#16)
Sodium hydroxide by-product   18.69MM
H_2 by-product used up in HCl manufacture

Adipic acid from imported cyclohexane, air ox.   (2.358) (#60)
  (85%) cyclohexane import 39.94MM

Refinery #4   (1.000) (#4)
Methane from H_2-C_1 stream, 11.98MM, uses 11.98/31.11 =
   0.385 of stream. Leftover for ammonia 0.615 of
   H_2-C_1 stream.

Ammonia for hexamethylene diamine (:H_2)   6.85MM
        for hydrogen cyanide   8.63MM
        for fertilizer   80.00MM
                                             95.48MM

Ammonia production from PLAT   43.91
   from leftover H_2C_1 43.06
   balance from C_2-C_2C_3-C_3 8.51
   which is 8.51/98.04 = 0.086801 of stream.

NH_3 from:	PLAT	0.615H_2C_1	0.087 C_2-C_2C_3-C_3	TOTAL		
H_2	35.95	5.50	0	41.45	(4.145)	(#31)
C_1	6.82	34.78	0	41.60	(4.160)	(#32)
$C_2^=$	0	1.28	2.22	3.50	(0.350)	(#33)
C_2	1.14	1.49	4.53	7.16	(0.716)	(#34)
$C_3^=$	0	0	1.03	1.03	(0.103)	(#35)
C_3	0	0	0.73	0.73	(0.073)	(#36)
	43.91	43.05	8.51	95.47		

Fertilizer: standard.   (8.768) (#55)
                                                     (6.896) (#56)
                                                     (6.688) (#43)

APPENDIX B

TABLE B-20
PRODUCTION PROGRAM: NYLON B
(mixed base)

Subprogram:	Staple	(3.650)	(#72)
	Salt	(3.650)	(#71)
	HMD	(1.872)	(#70)

<u>Adiponitrile</u> from tetrahydrofuran   (1.915) (#64)
<u>Tetrahydrofuran</u> from furfural   (1.679) (#62)

Furfural	28.43MM	not scheduled, since fur-
Bagasse	625MM	fural differential taken
Sulfuric acid	6.40MM	directly

<u>Sodium Cyanide</u>   18.46MM   from HCN and NaOH   (1.846) (#63)
<u>HCN</u>   10.15MM   from methane   11.98MM   (1.015) (#39)
                               and ammonia   8.63MM
<u>Sodium hydroxide</u> required,   15.14MM

<u>Hydrochloric acid</u> for adiponitrile, from chlorine   (1.703) (#17)
<u>Chlorine</u> by electrolysis   (1.661) (#16)
     Sodium hydroxide by-product   18.69MM
     Hydrogen by-product used up in H_2 manufacture

<u>Adipic acid</u> from imported cyclohexane, nitric acid ox.   (2.358) (#59)
     (85%) cyclohexane import   29.59MM
     (<u>Nitric acid</u> for the oxidation, 29.16MM)
     (<u>Ammonia</u>, for making this nitric acid, 8.34MM)

<u>Refinery</u> #4   (1.000) (#4)
<u>Methane</u> from H_2C_1 stream, 11.98MM, uses 11.98/31.11 =
     0.385 of stream. Leftover for ammonia production:
     0.615 of stream.

Ammonia needed:
     for hexamethylene diamine (H_2)   6.85MM
     for hydrogen cyanide   8.63MM
     for nitric acid (for oxidation)   8.34MM
     for fertilizer   80.00MM
                               Total   103.82MM

Production:
     from PLAT   43.91
     from leftover H_2C_1 43.06
     balance from C_2-C_2C_3-C_3 16.85,
     which is 16.85/98.04 = 0.172 of stream.

NH_3 from:	PLAT	0.615H_2C_1	0.172C_2-C_2C_3-C_3	TOTAL		
H_2	35.95	5.50	0	41.45	(4.145)	(#31)
C_1	6.82	34.78	0	41.60	(4.160)	(#32)
$C_2^=$	0	1.28	4.40	5.68	(0.568)	(#33)
C_2	1.14	1.49	8.98	11.61	(1.161)	(#34)
$C_3^=$	0	0	2.03	2.03	(0.203)	(#35)
C_3	0	0	1.44	1.44	(0.144)	(#36)
	43.91	43.05	16.85	103.81		

Fertilizer: standard   (8.768) (#55)
                                                            (6.896) (#56)

Nitric acid: for fertilizer   66.88MM
                    for oxidation   <u>29.16MM</u>
                                   96.04MM   (9.604) (#43)

TABLE B-21
PRODUCTION PROGRAM: NYLON C
(mixed base)

	Subprogram:	Staple Salt HMD	(3.650) (3.650) (1.872)	(#72) (#71) (#70)

<u>Adiponitrile</u> from tetrahydrofuran   (1.915) (#64)
<u>Tetrahydrofuran</u> from furfural   (1.679) (#62)

Furfural	28.43MM	not scheduled, since
Bagasse	625MM	furfural differential
Sulfuric acid	6.40MM	taken directly

<u>Sodium Cyanide</u>	18.46MM	from HCN and NaOH		(1.846)	(#63)
<u>HCN</u>	10.15MM	from methane	11.98MM	(1.015)	(#39)
		and ammonia	8.63MM		
<u>Sodium hydroxide</u> required,			15.14MM		

<u>Hydrochloric acid</u> for adiponitrile, from chlorine   (1.703) (#17)
<u>Chlorine</u> by electrolysis   (1.661) (#16)
    Sodium hydroxide by-product 18.69MM
    Hydrogen by-product used up in HCl manufacture.

<u>Adipic acid</u>, 23.58, from cyclohexane, nitric acid ox.   (2.358) (#59)
    (Nitric acid for the oxidation,   29.16MM)
    (Ammonia, for making this nitric acid,   8.34MM)
<u>Cyclohexane</u> (85% basis) from benzene, by hydrogenation   (2.916) (#58)
    Hydrogen needed   1.78MM
    equivalent ammonia   8.90MM

<u>Benzene</u> (from #4 refinery) 23.58MM needed
    (benzene process applied to B#4 refinery yields   (1.000) (#57)
    24.36.) Surplus is locally consumed.
<u>Refinery</u> #4   (1.000) (#4)
<u>Methane</u> from H_2C_1 stream, 11.98MM, uses 11.98/31.11 =
    0.385 of stream. Leftover for ammonia production,
    0.615 of stream.

<u>Ammonia</u> needed
    for hexamethylene diamine (:H_2)   6.85MM
    for hydrogen cyanide   8.63MM
    for nitric acid (oxidation)   8.34MM
    for cyclohexane (:H_2)   8.90MM
    for fertilizer   <u>80.00MM</u>
                                                  112.72MM

Production from PLAT   43.91
    from leftover H_2C_1 <u>43.06</u>
balance from C_2-C_2C_3=C_3 25.75 ,
    which is 25.75/98.04 = 0.263 of stream.

NH_3 from:	PLAT	0.615H_2C_1	0.263C_2-C_2C_3-C_3	TOTAL		
H_2	35.95	5.50	0	41.45	(4.145)	(#31)
C_1	6.82	34.78	0	41.60	(4.160)	(#32)
$C_2^=$	0	1.28	6.72	8.00	(0.800)	(#33)
C_2	1.14	1.49	13.72	16.35	(1.635)	(#34)
$C_3^=$	0	0	3.10	3.10	(0.310)	(#35)
C_3	0	0	2.21	2.21	(0.221)	(#36)
	43.91	43.05	25.75	112.71		

<u>Fertilizer</u>: standard   (8.768) (#55)
                                                            (6.896) (#56)

<u>Nitric acid</u>:   for fertilizer   66.88MM
              for oxidation   <u>29.16MM</u>
                               96.04MM   (9.604) (#43)
Note: Scale of benzene production very low.

APPENDIX B

TABLE B-22
PRODUCTION PROGRAM: NYLON D
(mixed base)

```
                                          Subprogram:   (3.650)  (#72)
                                                        (3.650)  (#71)
                                                        (1.872)  (#70)

Adiponitrile from imported butadiene   see Nylon F      (1.915)  (#67)

Adipic acid from imported cyclohexane, air ox.          (2.358)  (#60)
    (85%) cyclohexane import 39.94MM

Fertilizer: standard                                    (8.768)  (#55)
                                                        (6.896)  (#56)
                                                        (6.688)  (#43)

Refinery #4                                             (1.000)  (#4)
Methane for HCN, (see Nylon F) 12.43MM, uses 0.400 of
    stream; leftover for NH_3 0.600

Ammonia  for ADN, hydrogen:        :0.374MM H_2
         HMD hydrogenation         :1.369MM H_2
         credit: by-product H_2
         from electrolysis    -0.50
                              1.243MM
         H_2 equivalent NH_3   6.22MM
         for HCN               8.95MM
         for fertilizer       80.00MM
                              95.17MM

Production from PLAT           43.91
    from leftover H_2C_1       42.04
balance from C_2-C_2C_3-C_3     9.22,
    which is 0.940 of stream.
```

NH$_3$ from:	PLAT	0.60045 1H$_2$C$_1$	0.074969 C$_2$-C$_2$C$_3$-C$_3$	TOTAL		
H$_2$	35.95	5.37	0	41.32	(4.132)	(#31)
C$_1$	6.82	33.96	0	40.78	(4.078)	(#32)
C$_2^1$	0	1.25	2.41	3.66	(0.366)	(#33)
C$_2^2$	1.14	1.45	4.91	7.50	(0.750)	(#34)
C$_3^2$	0	0	1.11	1.11	(0.111)	(#35)
C$_3^3$	0	0	0.79	0.79	(0.079)	(#36)
	43.91	42.03	9.22	95.16		

TABLE B-23
PRODUCTION PROGRAM: NYLON E
(mixed base)

		Subprogram:	(3.650) (#72)
			(3.650) (#71)
			(1.872) (#70)

<u>Adiponitrile</u> from imported butadiene   see Nylon F   (1.915) (#67)

<u>Adipic acid</u> from imported cyclohexane, nitric acid ox.   (2.358) (#59)
    (85%) cyclohexane import   29.59MM
    (Nitric acid, for the oxidation   29.16MM)
    (Ammonia, for this nitric acid   8.34MM)

<u>Fertilizer</u>: standard   (8.768) (#55)
                                                                        (6.896) (#56)

<u>Nitric acid</u>: for fertilizer   66.88
              for oxidation   <u>29.16</u>
                      Total   96.04   (9.604) (#43)

<u>Refinery #4</u>   (1.000) (#4)
<u>Methane</u> for HCN, see Nylon F

<u>Ammonia</u>: Hydrogen for Adiponitrile   0.374MM
             for HMD hydrogenation   1.369MM H_2
             Credit: 0.50MM
             by-product H_2 from electrolysis
             Net 1.243MM; NH_3 equivalent = 6.22MM
               for HCN   8.95MM
               for nitric acid   8.34MM
               for fertilizer   <u>80.00MM</u>
                                         103.51MM

Production from PLAT   43.91
   from leftover H_2C_1 <u>42.04</u>
balance from C_2-C_2C_3-C_3 15.69

(See Nylon F)

NH_3 from:	PLAT	$0.600H_2C_1$	$0.179C_2$-C_2C_3-C_3	TOTAL		
H_2	35.95	5.37	0	41.32	(4.132)	(#31)
C_1	6.82	33.96	0	40.78	(4.078)	(#32)
$C_2^=$	0	1.25	4.59	5.84	(0.584)	(#33)
C_2	1.14	1.45	9.35	11.94	(1.194)	(#34)
$C_3^=$	0	0	2.12	2.12	(0.212)	(#35)
C_3	0	0	1.50	1.50	(0.150)	(#36)
	43.91	42.04	17.56	103.51		

APPENDIX B

TABLE B-24
PRODUCTION PROGRAM: NYLON F
(mixed base)

Subprogram	Staple	(3.650) (#72)
	Salt	(3.650) (#71)
	HMD	(1.872) (#70)

Adiponitrile from imported butadiene .... (1.915) (#67)
Butadiene import 14.17MM
Hydrogen Cyanide required, ... 10.53MM .... (1.053) (#39)
   from methane, ... 12.43MM
   and ammonia 8.95MM
Hydrogen required, ... 0.374MM
Chlorine required, ... 17.43MM (elec- ... (1.743) (#16)
                                                  trolysis)
   Sodium hydroxide by-product 19.61MM
   hydrogen by-product 0.50MM
Hydrochloric acid, by-product of adiponitrile, 1.92MM

Adipic acid, 23.58, from cyclohexane, nitric acid ox. .... (2.358) (#59)
   Nitric acid, for the oxidation, 29.16MM
   Ammonia, for making this nitric acid, 8.34MM
Cyclohexane (85% basis), from benzene by hydrogenation .... (2.916) (#58)
   Hydrogen needed 1.78MM
   equivalent ammonia 8.90MM
Benzene (from #4 refinery) 23.58MM needed (benzene .... (1.000) (#57)
   process applied to #4 refinery yields 24.36)
   Surplus is locally consumed.

Refinery #4 .... (1.000) (#4)
Methane from H_2C_1 stream, 12.43MM, uses 12.43/31.11=0.400
   of stream. Leftover for ammonia production, 0.600
   of stream.

Ammonia needed:
  1. for hydrogenations
      Adiponitrile   =   0.374
      HMD   =   1.369MM
      cyclohexane   =   1.780MM
      credit: electrolysis = .50 MM
                                    3.02 MM
      equivalent ammonia   15.10MM
  2. for hydrogen cyanide   8.95MM
  3. for nitric acid (oxidation) 8.34MM
  4. for fertilizer   80.00MM
                               112.39MM
Production from PLAT   43.91
   from leftover H_2C_1 42.04
balance from C_2-C_2C_3-C_3 26.45 ,
   which is 26.44/98.04 = 0.270 of stream.

NH_3 from:	PLAT	$0.600H_2C_1$	$0.270C_2$-C_2C_3-C_3	TOTAL		
H_2	35.95	5.37	0	41.32	(4.132)	(#31)
C_1	6.82	33.96	0	40.78	(4.078)	(#32)
$C_1^=$	0	1.25	6.90	8.15	(0.815)	(#33)
C_2	1.14	1.45	14.10	16.69	(1.669)	(#34)
$C_2^=$	0	0	3.18	3.18	(0.318)	(#35)
C_3	0	0	2.27	2.27	(0.227)	(#36)
	43.91	42.03	26.45	112.39		

Fertilizer: standard .... (8.768) (#55)
                                                                                                     (6.896) (#56)

Nitric acid: for fertilizer 66.88
           for oxidation   29.16
                              96.04 .... (9.604) (#43)
Note: Scale of benzene production is very low.

TABLE B-25
PRODUCTION PROGRAM: NYLON G

(All adipic acid base)
(imported cyclohexane)

			Subprogram:	staple salt HMD ADN	(3.650) (#72) (3.650) (#71) (1.872) (#70) (1.915) (#68)

<u>Adipic acid</u> from cyclohexane (air ox.) (5.845) (#59)
<u>Cyclohexane</u> import 99.01MM

<u>Refinery</u> #4 (1.000) (#4)

Ammonia: for fertilizer 80MM
          for ADN   6.51MM
          for HMD (:H_2) 6.85MM
                         93.36MM needed

Production from PLAT   43.91
Balance from H_2C_1 49.45, = 49.45/70.01 = 0.706
                        of H_2-C_1.

NH_3 from:	PLAT	$0.706 H_2C_1$	TOTAL	
H_2	35.95	6.32	42.27	(4.227) (#31)
C_1	6.82	39.93	46.75	(4.675) (#32)
$C_2^=$	0	1.47	1.47	(0.147) (#33)
C_2	1.14	1.71	2.85	(0.285) (#34)
	43.91	49.43	93.34	

<u>Fertilizer</u> (8.768) (#55)
               (6.896) (#56)
               (6.688) (#43)

TABLE B-26
PRODUCTION PROGRAM: NYLON H

(All Adipic acid base)
(Imported cyclohexane)

Subprogram: Staple	(3.650)	(#72)
Salt	(3.650)	(#71)
HMD	(1.872)	(#70)
ADN	(1.915)	(#68)

Adipic acid from cyclohexane by nitric acid ox. (5.845) (#59)
Cyclohexane import 73.35MM.
(Nitric acid 72.30MM, from ammonia 20.68)

Refinery #4   (1.000) (#4)

Ammonia: for fertilizer   80.00MM
          for ADN   6.51MM
          for HMD (:H_2)   6.85MM
                                93.36MM

          for nitric acid for ox. 20.68
                                114.04

Production from PLAT   43.91
     from H_2C_1 70.01
Balance from $C_2-C_2C_3-C_3$: 0.12 = 0.12/98.04 = 0.001 of stream.

NH_3 from:	PLAT	H_2-C_1	$0.001224C_2-C_2C_3-C_3$	TOTAL		
H_2	35.95	8.95	0.00	44.90	(4.490)	(#31)
C_1	6.82	56.56	0.00	63.38	(6.338)	(#32)
$C_2^=$	0	2.08	0.03	2.11	(0.211)	(#33)
C_2	1.14	2.42	0.06	3.62	(0.362)	(#34)
$C_3^=$	0	0	0.01	0.01	(0.001)	(#35)
C_3	0	0	0.01	0.01	(0.001)	(#36)
	43.91	70.01	0.11	114.03		

Fertilizer   (8.768) (#55)
                                                 (6.896) (#56)

Nitric acid: for ammonium nitrate 66.88MM
              for oxidation   72.30MM
                                   139.18MM   (13.918)(#43)

TABLE B-27
PRODUCTION PROGRAM: NYLON J

(All adiponitrile base, by way of furfural)	Subprogram	(3.650) (#72) (3.650) (#71) (1.872) (#70) (2.358) (#69)

<u>Hydrochloric acid</u> needed for adiponitrile, 34.25MM   (3.425) (#17)
<u>Chlorine</u> for HCl   33.40MM   (3.340) (#16)
   Sodium hydroxide by-product   37.60MM
   Hydrogen by-product used up for HCl.

<u>Sodium cyanide</u> for adiponitrile,   37.11MM   (3.711) (#63)
<u>Hydrogen cyanide</u> for sodium cyanide,   20.40   (2.040) (#39)
<u>Sodium hydroxide</u> for sodium cyanide,   30.43
   (by-product of chlorine electrolysis sufficient)

<u>Methane</u> for HCN needed 24.1MM. The H_2C_1 stream, available 31.11. Therefore, use 24.1/31.11 = 0.771 of H_2C_1 stream. Leftover 0.229 of H_2C_1 stream.

<u>Ammonia</u> needed for HCN   17.35
         for HMD (H_2 eq.)   6.85
         for fertilzer   80.00
                       <u>-6.06</u>   (see below)
                       98.14
Produced from PLAT   -43.91
    from leftover H_2C_1 <u>-16.01</u>
balance from $C_2-C_2C_3-C_3$ 38.22 ,
which is 38.22/98.04 = 0.390 of $C_2-C_2C_3-C_3$ stream.

NH_3 from:	PLAT	$0.229H_2C_1$	$0.390 C_2-C_2C_3-C_3$	TOTAL		
H_2	35.95	2.05	0	38.00	(3.800)	(#31)
C_1	6.82	12.95	0	19.77	(1.977)	(#32)
$C_2^=$	0	0.48	9.98	10.46	(1.046)	(#33)
C_2	1.14	0.55	20.37	22.06	(2.206)	(#34)
$C_3^=$	0	0	4.61	4.61	(0.461)	(#35)
C_3	0	0	3.28	3.28	(0.328)	(#36)
	43.91	16.03	38.24	98.18		

<u>Fertilizer</u>: total ammonia equivalent = 80.00MM. However, activity #69 has ammonium sulfate by-product 23.58MM, with an ammonia equivalent of 6.06MM. Therefore, ammonia needed for fertilizer = 80.00 - 6.06 = 73.94MM. 1/2 of this to ammo. nitrate,   (8.104) (#55)
1/2 to urea   (6.374) (#56)
nitric acid needed 61.81MM   (6.181) (#43)

<u>Refinery #4</u>   (1.000) (#4)
<u>Adiponitrile</u> from tetrahydrofuran   (3.851) (#64)
<u>Tetrahydrofuran</u> from furfural   (3.376) (#62)

    Furfural   57.17MM   not scheduled, since
    Bagasse   1257.0 MM   furfural differential
    Sulfuric acid   12.87MM   taken directly.

APPENDIX B

TABLE B-28
PRODUCTION PROGRAM: NYLON K

(All adiponitrile base by way of imported butadiene)	Subprogram	(3.650)	(#72)
		(3.650)	(#71)
		(1.872)	(#70)
		(2.358)	(#69)

Adiponitrile from butadiene. (HCl by-product 3.85MM.)   (3.851) (#67)
Butadiene import 28.51MM
Hydrogen cyanide from adiponitrile   21.18MM   (2.118) (#39)
    needed for HCN: methane   24.99MM
                      ammonia   18.00MM

Chlorine for adiponitrile   35.04MM   (3.504) (#16)
    Sodium hydroxide by-product   39.42MM
    Hydrogen by-product   1.02MM

Refinery #4   (1.000) (#4)
Methane needed for HCN 24.99; in H_2C_1 stream, available 31.11; therefore use 24.99/31.11 = 0.804 of H_2C_1 stream. Leftover = 0.196 of stream.

Ammonia needed
1. for hydrogenations: H_2 to ADN = 0.75; to HMD 1.37; credit for by-product H_2 from electrolysis, 1.02; net 1.10. Ammonia equivalent = 5.50
2. fertilizer   80.00
                -6.06   see below
3. for HCN   18.00
                97.44

Production from PLAT   -43.91
    from leftover H_2C_1 -13.72
balance from $C_2-C_2C_3=C_3$ 39.81,
which is 39.81/98.04 = 0.406 of $C_2-C_2C_3-C_3$ stream.

NH_3 from:	PLAT	0.196H_2C_1	0.406$C_2-C_2C_3-C_3$	TOTAL		
H_2	35.95	1.75	0	37.70	(3.770)	(#31)
C_1	6.82	11.09	0	17.91	(1.791)	(#32)
$C_2^=$	0	0.41	10.40	10.81	(1.081)	(#33)
C_2	1.14	0.47	21.21	22.82	(2.282)	(#34)
$C_3^=$	0	0	4.80	4.80	(0.480)	(#35)
C_3	0	0	3.41	3.41	(0.341)	(#36)
	43.91	13.72	39.82	97.45		

Fertilizer: see Nylon J   (8.104) (#55)
    (6.374) (#56)
    (6.181) (#43)

Appendix C

Estimated Production Costs: Chemical Intermediate Commodities

This appendix consists principally of the group of tables that contain estimates of mainland production costs of certain chemical intermediate materials. These materials are the ones that enter the production of the major fiber intermediates and fertilizer commodities. The data of the tables indicate the most efficient (i.e., cheapest) method of producing each required intermediate material. This information in turn makes possible the production-cost estimates used in Chapters 8 and 9 for the final fiber intermediates and fertilizer commodities—estimates that represent the lowest possible mainland costs of production at the output scales assumed. Strictly speaking, a complete analysis would estimate for every chemical product the costs of production via every possible method of manufacture. However, the nature of the processes involved and the broad character of the locational analysis make it possible to limit the number of cost computations as follows:

1. For any given product, multiple cost estimates are made only if there exist major process differences in the alternative production methods, i.e., only if basically different alternative raw materials, or significantly different types or proportions of process chemicals, can be used in its manufacture.

2. If one process for producing a given product has over a period of years shown itself to be economically superior and has become clearly dominant in the United States, no cost estimates for possible alternative processes are made in this appendix.

3. If different sets of costs have been estimated in the chemical

trade or professional literature for minor variations of a major productive process, the form of the process that results in the lowest production cost is the one chosen for presentation in this appendix.

Rates: Utilities and Labor

The unit rates at which utility inputs are charged are the same for all products and are as follows: steam, 40¢/M lb; electric power, 0.5¢/kw-hr; process fuel, 15¢/MM Btu; cooling water, 1.5¢/M gal. These rates are typical of those used in most cost estimates appearing in current chemical and petroleum trade literature and applying to the Gulf-Southwest area. It is assumed that, whatever the general scale of typical mainland operations may be, the scale of integrated steam and power-plant and cooling-water units will likely be at or near an optimum. Therefore, there are assumed to be no economies of scale differentials with respect to these utilities. Process fuel, whatever its source, tends to have its value set by that of the dominant fuel in the area: natural gas.

For most processes, the wage rate for direct labor is taken to be $2.50 per man-hour. This figure agrees generally with those used in other cost estimates appearing in the literature. For a very few inorganic-chemical processes, the wage rate used in the cost estimates is $2.25 per man-hour. This lower rate reflects the fact that less skilled labor is required for such processes.

Unit Rates: Basic Raw Materials

The rates for basic raw materials such as limestone, rocksalt, and coke are based on extensive lists of chemical and related prices which appear quarterly in *Chemical and Engineering News*. All basic hydrocarbon raw materials, such as ethane, propane, butane, and methane, are costed in accordance with their fuel value at a rate of 15¢/MM Btu. Process chemicals and catalyst materials which are used in small amounts are generally costed at rates given in the *Chemical and Engineering News* quarterly price lists.

Unit Rates: Ammonia and Ethylene

It will be noted that the lowest estimated costs for ammonia and ethylene are not the rates at which these materials are charged as inputs for other activities. This practice is explained in detail for ammonia in Chapter 8. In the case of ethylene, Table C–10 of this appendix shows that production from ethane is the cheaper process. Also it is probable that in the future an increasing proportion of ethylene production will be based on ethane from natural-gas stripping operations. However, such operations do not provide pure ethane streams, but rather a mixture of ethane, propane, and minor amounts of heavier hydrocarbons. Since ethylene operations based on refinery gas also utilize a mixed hydrocarbon raw material, it is reasonable to expect future ethylene-production cost to be closer to that of the refinery-gas process than to that of the pure ethane process.[1]

Method for Calculating Capital and Indirect Costs

The computation of all capital and indirect costs is based on a uniform set of percentages of plant investment cost and direct labor cost. As discussed in Chapter 3, the percentages adopted are basically the ones established by Walter Isard and Eugene W. Schooler in *Location Factors in the Petrochemical Industry,* U. S. Department of Commerce, Office of Technical Services, 1955. However, in the estimation of mainland petrochemical-production costs for the present study, two modifications of that set of percentages are made:

1. The item "indirect production cost" is calculated as equal to 40% (instead of 50%) of operating labor, supervision, plant maintenance, and equipment and operating supplies. This lower figure is deemed appropriate inasmuch as the operations considered are assumed typically to combine a number of individual activities, making possible integration savings in indirect costs.

2. The item "general office overhead" is not included. It is, in effect, an additional indirect production cost, applicable mostly to the operation of branch plants or units.

Sources of Data for the Preparation of Production-Cost Estimates

For the activities in general, a principal source of information on requirements of raw materials, utilities, and labor, as well as on investment-cost estimates, is Isard and Schooler (*op. cit.*). Ad-

---

[1] It may be observed that in acetylene production the production cost via ethane is used for subsequent calculations. The reason is that there is no appreciable increase in acetylene-production cost if the ethane is mixed with heavier hydrocarbons.

APPENDIX C

ditional information for a number of processes can be found in Faith, Keyes, and Clark, *Industrial Chemicals,* and in Shreve, *The Chemical Process Industries.* Useful data concerning estimated input requirements and investment costs for the production of ethylene and ethylene products appear in *The Manufacture of Ethylene and its Major Derivatives in Oklahoma,* a report by the Blaw-Knox Company prepared for the Oklahoma Planning and Resources Board.

All investment-cost estimates have been checked where possible by reference to reports of current plant construction in petroleum and chemical trade publications: e.g., the "Who's Building" sections of *Petroleum Refiner.* Sources of information on process descriptions, estimated production-cost elements, and estimated plant investment costs for several specific chemical processes are as follows:

Ammonia: B. J. Mayland, E. A. Comley, and J. C. Reynolds, "Ammonia Synthesis Gas," *Chemical Engineering Progress,* 50, 177–181 (1954); Barret S. Duff, "Economics of Ammonia Manufacture from Several Raw Materials," *Chemical Engineering Progress,* 51, 12-J to 16-J (1955); Carl Pfeiffer and Henry J. Sandler, "Ammonia from Catalytic Reformer Off-Gas," *Petroleum Refiner,* 34, 145–152 (May 1955)

Acetylene: Marcel J. P. Bogart and Robert H. Dodd, "Recent Developments in Wulff Acetylene," *Chemical Engineering Progress,* 50, 372–375 (1954); John Happel and Charles Marsel, "Processes Compete in Acetylene Boom," *Chemical Industries Week,* 68, 17–26 (1951); Marshall Sittig, "Acetylene—Challenge to the Petrochemical Industry," *Petroleum Processing,* 10, 1011–1023 (1955); T. P. Forbath and B. J. Gaffney, "Acetylene by the BASF Process," *Petroleum Refiner,* 33, 160–165 (December 1954).

Aromatics (Benzene): Davis Read, "The Production of High Purity Aromatics for Chemicals," *Petroleum Refiner,* 31, 97–103 (May 1952); and "An Appraisal of Catalytic Reforming," *Petroleum Processing,* 10, 1158–1204 (1955).

Butadiene: G. F. Hornaday, "Economics of Houdry Dehydrogenation," *Petroleum Refiner,* 33, 173–176 (Dec. 1954); "Butadiene from Normal Butane in One Step," *Oil and Gas Journal,* 52, 100–104 (Sept. 28, 1953); and Rubber Producing Facilities Disposal Commission, *Government-Owned Synthetic Rubber Facility, PLANCOR 706, Lake Charles, Louisiana,* PBD-1, Washington, 1953.

Chlorine: "Mercury Cell Chlorine and Caustic," *Industrial and Engineering Chemistry,* 45, 1824–1835 (1953); Bryce B. Schofield, "How Plant Costs Vary with Size," *Chemical Engineering,* 62, 185 (October 1955).

Urea: William H. Town Jr., "How the Competitive Urea Processes Compare Today," *Chemical Engineering,* 62, 186–190 (October 1955); L. H. Cook, "Urea," *Chemical Engineering Progress,* 50, 327–331 (1954).

TABLE C-1
PRODUCTION COST OF AMMONIA, $/100 LB.

	From Fuel Oil via Partial Oxidation			From Catalytic Reformer Off-Gas (85.4% hydrogen)			From Natural or Refinery Gas Hydrocarbons via Partial Oxidation			From Coal via Partial Oxidation	
Plant Size: (MM lb / yr)	65	100	135	65	100	135	65	100	135	65	135
Plant Investment: (MM $)	3.750	5.190	6.500	3.300	4.565	5.750	3.600	4.980	6.250	4.000	6.750
Raw Materials:											
Oil Feed, @15¢/MMBtu	0.205	0.205	0.205								
Reformer Gas Feed, @15¢/MMBtu				0.178	0.178	0.178					
Hydrocarbon Feed, @15¢/MMBtu							0.208	0.208	0.208		
Coal Feed, @15¢/MMBtu										0.242	0.242
Shift Catalyst, @72¢/lb	0.013	0.013	0.013	0.003	0.003	0.003	0.011	0.011	0.011	0.013	0.013
Synthesis Catalyst, @60¢/lb	0.015	0.015	0.015	0.013	0.013	0.013	0.015	0.015	0.015	0.015	0.015
Caustic, @2.7¢/lb	0.011	0.011	0.011	0.004	0.004	0.004	0.011	0.011	0.011	0.011	0.011
Monoethanolamine, @25¢/lb	0.005	0.005	0.005	0.024	0.024	0.024	0.003	0.003	0.003	0.005	0.005
Lube Oil, @80¢/gal	0.020	0.020	0.020				0.020	0.020	0.020	0.020	0.020
Utilities:											
Fuel, @15¢/MMBtu	0.024	0.024	0.024	0.075	0.075	0.075	0.024	0.024	0.024	0.024	0.024
Electric Power, @ 0.5¢/kwh	0.225	0.225	0.225	0.009	0.009	0.009	0.222	0.222	0.222	0.225	0.225
Steam, @40¢/M lb	0.075	0.075	0.075	0.134	0.134	0.134	0.070	0.070	0.070	0.075	0.075
Cooling Water, @1.5¢/M gal	0.041	0.041	0.041	0.072	0.072	0.072	0.041	0.041	0.041	0.041	0.041
Direct Labor: @ 2.50/hr	0.210	0.158	0.105	0.210	0.157	0.105	0.210	0.157	0.105	0.210	0.105
Capital and Indirect Costs:											
Supervision	0.021	0.016	0.011	0.021	0.016	0.011	0.021	0.016	0.011	0.021	0.011
Plant maintenance	0.227	0.208	0.197	0.200	0.182	0.174	0.218	0.199	0.189	0.242	0.204
Equipment and Operating Supplies	0.034	0.031	0.030	0.030	0.027	0.026	0.033	0.030	0.028	0.036	0.031
Payroll Overhead	0.052	0.041	0.032	0.049	0.040	0.030	0.051	0.041	0.032	0.053	0.033
Indirect Production Cost	0.197	0.164	0.137	0.184	0.153	0.126	0.193	0.161	0.133	0.204	0.141
Depreciation	0.568	0.519	0.492	0.500	0.457	0.435	0.545	0.498	0.474	0.606	0.511
Taxes	0.057	0.052	0.049	0.050	0.046	0.044	0.055	0.050	0.047	0.061	0.051
Insurance	0.057	0.052	0.049	0.050	0.046	0.044	0.055	0.050	0.047	0.061	0.051
Interest	0.227	0.208	0.197	0.200	0.182	0.174	0.218	0.199	0.189	0.242	0.204
	2.284	2.083	1.933	2.006	1.818	1.681	2.224	2.026	1.880	2.407	2.013

TABLE C-2
PRODUCTION COST OF NITRIC ACID, $/100 LB.

	40	60	70	100	140	200
Plant Size: (MM lb/yr.)						
Plant Investment: (MM $)	3.256	4.205	4.633	5.800	7.600	10.100
Raw Materials:						
Ammonia, @ 2.033¢/lb	0.581	0.553	0.553	(1) 0.553	(1) 0.553	0.538
@ 1.933¢/lb				(2) 0.538	(2) 0.538	
@ 1.880¢/lb						
Utilities:						
Cooling Water, @ 1.5¢/M gal	0.012	0.012	0.012	0.012	0.012	0.012
Electric Power, @ 0.5¢/kwh	0.060	0.060	0.060	0.060	0.060	0.060
Direct Labor: @ $2.50/hr	0.180	0.130	0.115	0.088	0.067	0.053
Capital and Indirect Costs:						
Supervision	0.018	0.013	0.012	0.009	0.007	0.005
Plant Maintenance	0.326	0.280	0.265	0.232	0.217	0.202
Equipment and Operating Supplies	0.049	0.042	0.040	0.035	0.033	0.030
Payroll Overhead	0.054	0.042	0.039	0.032	0.027	0.024
Indirect Production Cost	0.229	0.186	0.173	0.146	0.130	0.116
Depreciation	0.814	0.700	0.663	0.580	0.543	0.505
Taxes	0.081	0.070	0.066	0.058	0.054	0.051
Insurance	0.081	0.070	0.066	0.058	0.054	0.051
Interest	0.326	0.280	0.265	0.232	0.217	0.202
Total	2.811	2.438	2.329	(1) 2.095	(1) 1.974	1.849
				(2) 2.080	(2) 1.959	

TABLE C-3
PRODUCTION COST OF AMMONIUM NITRATE, $/100 LB.

		50	75	90	100	200
Plant Size: (MM lb/yr)						
Plant Investment: (MM $)		0.380	0.500	0.566	0.608	0.975
Raw Materials:						
Ammonia,	@ 2.033¢/lb	0.484				
	@ 1.933¢/lb		0.460			
	@ 1.880¢/lb			0.460	(1) 0.460	0.447
	@ 2.811¢/lb				(2&3) 0.447	
Nitric Acid,	@ 2.438¢/lb	2.144				
	@ 2.329¢/lb		1.920			
	@ 2.095¢/lb			(1) 1.777		
	@ 1.974¢/lb			(2) 1.598	(1) 1.598	
	@ 2.080¢/lb			(3) 1.506		
	@ 1.959¢/lb				(2) 1.587	1.412
	@ 1.849¢/lb				(3) 1.495	
Utilities:						
Steam,	@ 40¢/M lb	0.026	0.026	0.026	0.026	0.026
Cooling Water,	@ 1.5¢/M gal	0.001	0.001	0.001	0.001	0.001
Electric Power,	@ 0.5¢/kwh	0.008	0.008	0.008	0.008	0.008
Direct Labor:						
	@ $2.50/mhr	0.155	0.115	0.103	0.095	0.058
Capital and Indirect Costs:						
Supervision		0.016	0.012	0.010	0.010	0.006
Plant Maintenance		0.030	0.027	0.025	0.024	0.020
Equipment and Operating Supplies		0.005	0.004	0.004	0.004	0.003
Payroll Overhead		0.028	0.021	0.019	0.018	0.011
Indirect Production Cost		0.082	0.063	0.057	0.053	0.035
Depreciation		0.075	0.058	0.063	0.060	0.050
Taxes		0.008	0.007	0.006	0.006	0.005
Insurance		0.008	0.007	0.006	0.006	0.005
Interest		0.030	0.027	0.025	0.024	0.020
Total		3.100	2.766	(1) 2.590	(1) 2.393	2.107
				(2) 2.411	(2) 2.369	
				(3) 2.319	(3) 2.277	

TABLE C-4
PRODUCTION COST OF UREA, $/100 LB.

		40	60	70	100	200
Plant Size: (MM lb./yr)						
Plant Investment: (MM $)		1.816	2.383	2.642	3.355	5.840
Raw Materials:						
Ammonia,	@ 2.033¢/lb.	1.179				
	@ 1.933¢/lb.		1.121	1.121	(1) 1.121	
	@ 1.880¢/lb.				(2) 1.090	1.090
Utilities:						
Steam,	@ 40¢/M lb.	0.110	0.110	0.110	0.110	0.110
Cooling Water,	@ 1.5¢/M gal.	0.036	0.036	0.036	0.036	0.036
Electric Power,	@ 0.5¢/kwh.	0.017	0.017	0.017	0.017	0.017
Fuel,	@ 15¢/MM Btu.	0.034	0.034	0.034	0.034	0.034
Direct Labor:	@ $2.50/hr.	0.270	0.195	0.173	0.130	0.080
Capital and Indirect Costs:						
Supervision		0.027	0.020	0.017	0.013	0.008
Plant Maintenance		0.182	0.159	0.151	0.134	0.117
Equipment and Operating Supplies		0.027	0.024	0.023	0.020	0.018
Payroll Overhead		0.058	0.043	0.040	0.032	0.022
Indirect Production Cost		0.202	0.159	0.146	0.119	0.098
Depreciation		0.455	0.398	0.378	0.335	0.293
Taxes		0.046	0.040	0.038	0.034	0.029
Insurance		0.046	0.040	0.038	0.034	0.029
Interest		0.182	0.159	0.151	0.134	0.117
Total		2.871	2.555	2.473	(1) 2.303	2.098
					(2) 2.272	

270 INDUSTRIAL COMPLEX ANALYSIS

TABLE C-5
PRODUCTION COST OF ACETYLENE, $/100 LB.

		From Ethane, via Wulff Process, Recycle Operation				From Natural Gas, via Wulff Process, Once-through Operation			From Natural Gas via Sachsse Process		From Calcium Carbide
Plant Size: (MM lb/yr)		20	25	50	100	200	25	100	200	100	200	100
Plant Investment: (MM $)		2.970	3.400	5.150	7.810	11.840	4.000	9.200	13.950	12.600	19.100	1.000
Raw Materials:												
Ethane,	@ 15¢/MMBtu	0.742	0.742	0.742	0.742	0.742						
Natural Gas,	@ 15¢/MMBtu						1.888	1.888	1.888	1.730	1.730	
Calcium Carbide	@3.113¢/lb											9.728
Solvent,	@ 33¢/lb	0.099	0.099	0.099	0.099	0.099	0.099	0.099	0.099			
	@ 40¢/lb									0.400	0.400	
Caustic,	@2.7¢/lb									0.095	0.095	
Lube Oil,	@ 50¢/gal									0.025	0.025	
Utilities:												
Steam,	@ 40¢/Mlb	0.645	0.645	0.645	0.645	0.645	1.046	1.046	1.046	0.520	0.520	
Electric Power	@ 0.5¢/kwh	0.018	0.018	0.018	0.018	0.018	0.032	0.032	0.032	0.130	0.130	-0.010
Cooling Water,	@ 1.5¢/Mgal	0.083	0.083	0.083	0.083	0.083	0.171	0.171	0.171	0.206	0.206	
Fuel Gas,	@ 15¢/MMBtu	-0.223	-0.223	-0.223	-0.223	-0.223	-1.295	-1.295	-1.295	-0.860	-0.860	
Oils and Tars,	@ 15¢/MMBtu	-0.013	-0.013	-0.013	-0.013	-0.013	-0.059	-0.059	-0.059			
Direct Labor:	@$2.50/mhr	0.449	0.396	0.235	0.139	0.089	0.396	0.139	0.089	0.139	0.089	0.148
Capital and Indirect Costs:												
Supervision		0.045	0.040	0.024	0.014	0.009	0.040	0.014	0.009	0.014	0.009	0.015
Plant Maintenance		0.594	0.544	0.412	0.312	0.237	0.640	0.368	0.279	0.504	0.382	0.042
Equipment and Operating Supplies		0.089	0.082	0.062	0.047	0.036	0.096	0.055	0.042	0.076	0.057	0.006
Payroll Overhead		0.119	0.106	0.070	0.046	0.032	0.113	0.051	0.036	0.293	0.215	0.084
Indirect Production Cost		0.471	0.425	0.293	0.205	0.148	0.469	0.230	0.168	0.061	0.043	0.028
Depreciation		1.485	1.360	1.030	0.780	0.593	1.600	0.920	0.698	1.260	0.955	0.105
Taxes		0.149	0.136	0.103	0.078	0.059	0.160	0.092	0.070	0.126	0.096	0.011
Insurance		0.149	0.136	0.103	0.078	0.059	0.160	0.092	0.070	0.126	0.096	0.011
Interest		0.594	0.544	0.412	0.312	0.237	0.640	0.368	0.279	0.504	0.382	0.042
Total		5.495	5.120	4.095	3.362	2.850	6.196	4.211	3.622	5.349	4.570	10.230

APPENDIX C

TABLE C-6
PRODUCTION COST OF AROMATICS, $/BBL.

Plant Size: (BSD feedstock charged)		3,350
(BSD aromatics produced)		
(Benzene)		290
(Toluene)		762
(Xylenes)		445
Plant Investment: (MM $) (for Udex unit)		1.345

Raw Material:
Feedstock, 3350 BSD depentanized platformate, clear octane 85.5, @ 10.5¢/gal. $14,773.50
Less Raffinate: 1853 BSD clear octane 66.4, @ 9.25¢/gal.   7,198.90
Net Feedstock Cost/SD $ 7,574.60
Net Feedstock Cost/bbl aromatics $5.060

Chemicals:
Solvent   0.024
Clay   0.060

Utilities:
Fuel Gas, @ 15¢/MM Btu   0.005
Cooling Water, @ 1.5¢/M gal   0.058
Electric Power, @ .5¢/kwh   0.001
Steam, @ 40¢/M lb   0.545

Royalty (2%):
Benzene   (1) 0.319
Toluene   (2) 0.286
Xylene   (3) 0.286

Direct Labor:
@ $2.50/mhr   0.080

Capital and Indirect Costs:

Supervision		0.008	
Plant Maintenance		0.109	
Equipment and Operating Supplies		0.016	
Payroll Overhead		0.021	
Indirect Production Cost		0.085	
Depreciation		0.272	
Taxes		0.027	
Insurance		0.027	
Interest		0.109	**$/gal.**
Total: Benzene	(1)	6.826	0.163
Toluene	(2)	6.793	0.162
Xylene	(3)	6.793	0.162

TABLE C-7
PRODUCTION COST OF BUTADIENE, $/100 LB.

Plant Size: (MM lb/yr)	53	100
Plant Investment: (MM $)	9	14
Raw Materials:		
Butane, @ .328¢/lb (Btu value)	0.623	0.623
Utilities:		
Steam, @ 40¢/lb	1.840	1.840
Electric Power, @ .5¢/kwh	0.081	0.081
Fuel, .5¢/MM Btu	0.497	0.497
Direct Labor:		
@ 2.50/mhr	0.300	
@ 2.50/mhr		0.200
Capital and Indirect Costs:		
Supervision	0.030	0.020
Plant Maintenance	0.679	0.560
Equipment and Operating Supplies	0.102	0.084
Payroll Overhead	0.101	0.075
Indirect Production Cost	0.444	0.346
Depreciation	1.698	1.400
Taxes	0.170	0.140
Insurance	0.170	0.140
Interest	0.679	0.560
Total	7.414	6.566

TABLE C-8
PRODUCTION COST OF CALCIUM CARBIDE, $/TON

Plant Size: (tons/yr)	150,000
Plant Investment: (MM $)	$10.1
Raw Materials:	
Lime, @ $11.40/ton	10.830
Coke, @ $15/ton	9.000
Electrodes, @ 10¢/lb	4.000
Utilities:	
Electric Power, @ .5¢/kwh	14.500
Direct Labor:	
@ 2.50/mhr	4.450
Capital and Indirect Costs:	
Supervision	0.445
Plant Maintenance	2.700
Equipment and Operating Supplies	0.405
Payroll Overhead	0.937
Indirect Production Cost	3.200
Depreciation	6.750
Taxes	0.675
Insurance	0.675
Interest	2.700
Total	62.267
(Total per 100 lb)	(3.113)

TABLE C-9
PRODUCTION COST OF CHLORINE GAS, $/100 LB.*

		20	30	66
Plant Size: (MM lb/yr)				
Plant Investment: (MM $)		4.085	5.535	10.0
Raw Materials:				
Rock Salt,	@ $9.00/ton	0.765	0.765	0.765
Mercury,	@ $3.73/lb	0.104	0.104	0.104
Graphite,	@ $0.45/lb	0.135	0.135	0.135
HCl,	@ $0.03/lb	0.060	0.060	0.060
NaOH,	@ $0.027/lb	0.027	0.027	0.027
				1.091
Utilities:				
Steam,	@ $0.40/M lb	0.020	0.020	0.020
Electric Power,	@ 0.5¢/kwh	0.810	0.810	0.810
Cooling Water,	@1.5¢/M gal	0.001	0.001	0.001
				0.831
Direct Labor:	@ $2.25/mhr	1.053	0.761	0.473
Capital and Indirect Costs:				
Supervision		0.105	0.076	0.047
Plant Maintenance		0.408	0.369	0.303
Equipment and Operating Supplies		0.123	0.111	0.091
Payroll Overhead		0.235	0.181	0.123
Indirect Production Cost		0.839	0.674	0.487
Depreciation		2.044	1.845	1.515
Taxes		0.204	0.185	0.152
Insurance		0.204	0.185	0.152
Interest		0.408	0.369	0.303
Total		7.545	6.678	5.568
Less Credit for NaOh	@ 2.7¢/lb	3.979	3.979	3.979
		3.566	2.699	1.589

*112 lb of Caustic Soda (NaOH) is produced with every 100 lb chlorine gas.

TABLE C-10
PRODUCTION COST OF ETHYLENE, $/100 LB.

		From Ethane				From Mixed Refinery Gas		
Plant Size: (MM lb/yr)		10	30	50	200	50	100	200
Plant Investment: (MM $)		1.26	2.64	3.70	9.40	3.27	5.20	8.27
Raw Materials:								
Ethane,	@ 15¢/MMBtu	0.446	0.446	0.446	0.446			
Mixed Refinery Gas,	@ 15¢/MMBtu					0.676	0.676	0.676
Utilities:								
Fuel,	@ 15¢/MMBtu	0.032	0.032	0.032	0.032	0.155	0.155	0.155
Steam,	@ 40¢/M lb	0.103	0.103	0.103	0.103	0.100	0.100	0.100
Cooling Water,	@ 1.5¢/M gal	0.073	0.073	0.073	0.073	0.027	0.027	0.027
Electric Power,	@ 0.5¢/kwh	0.011	0.011	0.011	0.011	0.225	0.225	0.225
Direct Labor:	@ $2.50/mhr	0.518	0.215	0.143	0.047	0.140	0.080	0.045
Capital and Indirect Costs:								
Supervision		0.052	0.021	0.014	0.005	0.014	0.008	0.005
Plant Maintenance		0.504	0.352	0.296	0.188	0.261	0.208	0.165
Equipment and Operating Supplies		0.076	0.053	0.044	0.028	0.039	0.031	0.025
Payroll Overhead		0.123	0.062	0.046	0.022	0.042	0.029	0.020
Indirect Production Cost		0.460	0.256	0.199	0.107	0.182	0.131	0.100
Depreciation		1.260	0.880	0.740	0.470	0.652	0.520	0.414
Taxes		0.126	0.088	0.074	0.047	0.065	0.052	0.041
Insurance		0.126	0.088	0.074	0.047	0.065	0.052	0.041
Interest		0.504	0.352	0.296	0.188	0.261	0.208	0.165
Total		4.414	3.032	2.591	1.814	2.904	2.502	2.204

TABLE C-11
PRODUCTION COST OF ETHYLENE DICHLORIDE, $/100 LB.

Plant Size: (MM lb/yr)			40	50	100
Plant Investment: (MM $)			$7.3	8.5	$13.8
Raw Materials:					
Ethylene,	@ 2.2¢/lb				0.649
	@ 3.1¢/lb		0.914		
	@ 2.9¢/lb			0.855	
Chlorine,	@ 1.589¢/lb	(1)	1.146	1.146	1.146
	@ 2.699¢/lb	(2)	1.946		
Utilities:					
Steam,	@ 40¢/M lb		0.004	0.004	0.004
Electric Power,	@ .5¢/kwh		0.001	0.001	0.001
Cooling Water,	@ 1.5¢/Mgal.		0.007	0.007	0.007
Direct Labor:					
	@ $2.50/mhr		0.059	0.052	0.030
Capital and Indirect Costs:					
Supervision			0.006	0.005	0.003
Plant Maintenance			0.730	0.680	0.552
Equipment and Operating Supplies			0.110	0.102	0.083
Payroll Overhead			0.065	0.060	0.046
Indirect Production Cost			0.362	0.336	0.267
Depreciation			1.825	1.700	1.380
Taxes			0.183	0.170	0.138
Insurance			0.183	0.170	0.138
Interest			0.730	0.680	0.552
Total		(1)	6.325	5.968	4.996
		(2)	7.125		

276 INDUSTRIAL COMPLEX ANALYSIS

TABLE C-12
PRODUCTION COST OF ETHYLENE OXIDE, $/100 LB.

		Via Chlorhydrin Process			Via Oxidation Process		
Plant Size: (MM lb/yr)		20	40	50	100	20	40	50	100
Plant Investment: (MM $)		3.163	4.878	5.620	8.648	3.946	6.086	7.000	10.780
Raw Materials:									
Ethylene,	@ 3.1¢/lb	3.317							
	@ 2.9¢/lb		3.103	3.103					
	@ 2.2¢/lb				2.354				
	@ 1.59¢/lb					3.658			
Chlorine,	@ 0.740¢/lb	3.816	3.816	3.816	3.816		3.422	3.422	2.596
Quicklime,	@ 0.586¢/lb	(1) 1.413							
	@ 0.570¢/lb	(2) 1.089	(1) 1.119	1.089	1.089				
Caustic Soda,	@ 2.7¢/lb	0.062	(2) 1.089	0.062	0.062				
Sulfuric Acid,	@ 0.95¢/lb	0.025	0.062	0.025	0.025				
Catalyst,	@ 1.25/lb		0.025			0.038	0.038	0.038	0.038
Utilities:									
Steam,	@ 40¢/Mlb	0.255	0.255	0.255	0.255	0.314	0.314	0.314	0.314
Cooling Water,	@ 1.5¢/Mgal	0.049	0.049	0.049	0.049	0.074	0.074	0.074	0.074
Electric Power,	@ 0.5¢/kwh	0.030	0.030	0.030	0.030	0.031	0.031	0.031	0.031
Fuel,	@ 15¢/MMBtu					0.043	0.043	0.043	0.043
Direct Labor:	@ $2.50/mhr	0.590	0.345	0.290	0.168	0.666	0.388	0.325	0.190
Capital and Indirect Costs:									
Supervision		0.059	0.035	0.029	0.017	0.067	0.039	0.033	0.019
Plant Maintenance		0.633	0.488	0.450	0.346	0.789	0.609	0.560	0.431
Equipment and Operating Supplies		0.095	0.073	0.068	0.052	0.118	0.091	0.084	0.065
Indirect Production Cost		0.551	0.376	0.335	0.233	0.169	0.110	0.096	0.064
Payroll Overhead		0.145	0.094	0.082	0.054	0.656	0.451	0.401	0.282
Depreciation		1.583	1.220	1.125	0.865	1.973	1.523	1.400	1.078
Taxes		0.158	0.122	0.113	0.087	0.197	0.152	0.140	0.108
Insurance		0.158	0.122	0.113	0.087	0.197	0.152	0.140	0.108
Interest		0.633	0.488	0.450	0.346	0.789	0.609	0.560	0.431
Total		(1) 13.572 (2) 13.248	(1) 11.822 (2) 11.792	11.484	9.935	9.779	8.046	7.661	5.872
Less By-products @ 7¢/lb		2.100	2.100	2.100	2.100				
Final Total		(1) 11.472 (2) 11.148	(1) 9.722 (2) 9.692	9.384	7.835	9.779	8.046	7.661	5.872

TABLE C-13
PRODUCTION COST OF HYDROGEN CHLORIDE FROM CHLORINE AND HYDROGEN, $/100 LB. 100% HCl

		20	35	100
Plant Size: (MM lb/yr)		20	35	100
Plant Investment: (MM $)		$1.265	1.86	$3.9
Raw Materials:				
Chlorine,	@ 3.566¢/lb	3.459		
	@ 1.589¢/lb			1.541
	@ 2.5¢/lb		2.425	
Hydrogen,	@ 15¢/MMBtu	0.037	0.037	0.037
Utilities:				
Steam,	@ 40¢/M lb	0.006	0.006	0.006
Electric Power,	@ .5¢/kwh	0.002	0.002	0.002
Cooling Water,	@ 1.5¢/M gal	0.510	0.510	0.510
Direct Labor:				
	@ 2.25/mhr	0.511		
	@ 2.25/mhr			0.141
	@ 2.25/mhr		0.326	
Capital and Indirect Costs:				
Supervision		0.051	0.033	0.014
Plant Maintenance		0.253	0.213	0.156
Equipment and Operating Supplies		0.038	0.032	0.023
Payroll Overhead		0.103	0.070	0.035
Indirect Production Cost		0.341	0.242	0.134
Depreciation		0.633	0.533	0.390
Taxes		0.063	0.053	0.039
Insurance		0.063	0.053	0.039
Interest		0.253	0.213	0.156
	Total	6.323	4.748	3.223

TABLE C-14
PRODUCTION COST OF HYDROGEN CYANIDE, $/100 LB.

Plant Size: (MM lb/yr)		10	20	25	50	100
Plant Investment: (MM $)		0.780	1.300	1.500	2.440	3.900
Raw Materials:						
Ammonia,	@ 2.033¢/lb		(1) 1.728	(1) 1.728		
	@ 1.933¢/lb	1.643	(2) 1.643	(2) 1.643	1.643	1.598
Natural Gas (methane)	@ 1.880¢/lb @ 15¢/MMBtu	0.420	0.420	0.420	0.420	0.420
Catalyst and Chemicals		0.200	0.200	0.200	0.200	0.200
Utilities:						
Steam,	@ 40¢/M lb	0.075	0.075	0.075	0.075	0.075
Electric Power,	@ 0.5¢/kwh	0.070	0.070	0.070	0.070	0.070
Cooling Water,	@ 1.5¢/M gal	0.084	0.084	0.084	0.084	0.084
Direct Labor:	@ $2.50/mhr	0.338	0.191	0.160	0.091	0.054
Capital and Indirect Costs:						
Supervision		0.034	0.019	0.016	0.009	0.005
Plant Maintenance		0.312	0.260	0.240	0.195	0.156
Equipment and Operating Supplies		0.047	0.039	0.036	0.029	0.023
Payroll Overhead		0.079	0.051	0.044	0.020	0.021
Indirect Production Cost		0.292	0.204	0.181	0.130	0.095
Depreciation		0.780	0.650	0.600	0.488	0.390
Taxes		0.078	0.065	0.060	0.049	0.039
Insurance		0.078	0.065	0.060	0.049	0.039
Interest		0.312	0.260	0.240	0.195	0.156
Total		4.842	(1) 4.381	(1) 4.214	3.747	3.425
			(2) 4.296	(2) 4.129		

TABLE C-15
PRODUCTION COST OF METHANOL, $/100 LB.

Plant Size: (MM lb/yr)	120
Plant Investment: (MM $)	9.720
Raw Materials:	
Natural Gas, @ 15¢/Mcf	0.339
Catalyst and Chemicals	0.017
Utilities:	
Steam, @ 40¢/Mlb.	0.040
Cooling Water, @ 1.5¢/M gal	0.063
Electric Power, @ .5¢/kwh	0.185
Direct Labor:	
@ $2.50/mhr	0.153
Capital and Indirect Costs	
Supervision	0.015
Plant Maintenance	0.324
Equipment and Operating Supplies	0.049
Payroll Overhead	0.050
Indirect Production Cost	0.216
Depreciation	0.810
Taxes	0.081
Insurance	0.081
Interest	0.324
Total	$2.747
($/gal.)	(0.181)

TABLE C-16
SEPARATION COST OF PARAXYLENE, $/GAL.

Plant Size:	(MM gal/yr, mixed xylene throughput)	34
	(MM gal/yr, paraxylene)	3.25
Plant Investment:	(MM $)	1.5
	(paraxylene separation unit)	

Utilities:
Steam, @ 40¢/M lb	0.00193
Electric Power, @ .5¢/kwh	0.00602
Cooling Water, @ 1.5¢/M gal.	0.00119

Direct Labor:
@ $2.50/mhr	0.01205

Capital and Indirect Costs:
Supervision	0.00121
Plant Maintenance	0.01825
Equipment and Operating Supplies	0.00274
Payroll Overhead	0.00336
Indirect Production Cost	0.01370
Depreciation	0.04563
Taxes	0.00456
Insurance	0.00456
Interest	0.01825
Total	0.13345
Transfer Cost of Xylene Raw Material	0.162
Total Transfer Cost of Paraxylene	$0.295
(per pound)	(.040)

APPENDIX C

TABLE C-17
PRODUCTION COST OF QUICKLIME, $/100 LB.

Plant Size: (MM lb/yr)	20	40	80	100
Plant Investment: (MM $)	0.085	0.138	0.2245	0.2625
Raw Materials:				
Limestone, @ $3.00/ton	0.282	0.282	0.282	0.282
Utilities:				
Fuel, @ 15¢/MM Btu	0.061	0.061	0.061	0.061
Direct Labor:				
@ $2.25/mhr	0.322	0.185	0.106	0.090
Capital and Indirect Costs:				
Supervision	0.032	0.019	0.011	0.009
Plant Maintenance	0.017	0.014	0.011	0.011
Equipment and Operating Supplies	0.003	0.002	0.002	0.002
Payroll Overhead	0.054	0.032	0.018	0.016
Indirect Production Cost	0.150	0.088	0.052	0.056
Depreciation	0.043	0.035	0.026	0.026
Taxes	0.004	0.004	0.003	0.003
Insurance	0.004	0.004	0.003	0.003
Interest	0.017	0.014	0.011	0.011
Total	0.989	0.740	0.586	0.570

TABLE C-18
PRODUCTION COST OF SODIUM CYANIDE, $/100 LB.

Plant Size (MM lb/yr):	20	40	100
Plant Investment (MM $):	1.275	2.000	3.900
Raw Materials:			
Hydrogen Cyanide @ 4.842¢/lb	2.663		
@ 4.296¢/lb		2.363	
@ 3.425¢/lb			1.887
Caustic Soda @ 2.700¢/lb	2.210	2.210	2.210
Utilities:			
Steam, @ 40¢/M lb	0.060	0.060	0.060
Electric Power, @ 0.5¢/kwh	0.015	0.015	0.015
Direct Labor:			
@$2.50/mhr	0.200	0.113	0.054
Capital and Indirect Costs:			
Supervision	0.020	0.011	0.005
Plant Maintenance	0.255	0.200	0.156
Equipment and Operating Supplies	0.038	0.030	0.023
Indirect Production Cost	0.205	0.142	0.095
Payroll Overhead	0.052	0.034	0.021
Depreciation	0.638	0.500	0.390
Taxes	0.064	0.050	0.039
Insurance	0.064	0.050	0.039
Interest	0.255	0.200	0.156
Total	6.739	5.978	5.150

TABLE C-19
PRODUCTION COST OF SULFURIC ACID, $/100 LB.

		20	100
Plant Size: (MM lb/yr)		20	100
Plant Investment: (MM $)		0.205	0.810
Raw Materials:			
Sulfur,	@ .0114$/lb	0.392	0.392
Utilities:			
Steam,	@ 40¢/M lb	0.036	0.036
Electric Power,	@ .5¢/kwh	0.002	0.002
Direct Labor:			
	@ $2.50/mhr	0.110	
	@ 2.50/mhr		0.042
Capital and Indirect Costs:			
Supervision		0.011	0.004
Plant Maintenance		0.041	0.032
Equipment and Operating Supplies		0.006	0.005
Payroll Overhead		0.021	0.009
Indirect Production Cost		0.067	0.033
Depreciation		0.103	0.080
Taxes		0.010	0.008
Insurance		0.010	0.008
Interest		0.041	0.032
	Total	0.850	0.683

Index

Acetone, as an input of the full programs, 70n
  transport cost of, 111
Acetylene, as a basic petrochemical for the full programs, 69, 79
  in Dynel programs, 186–187, 192, 200
  in Orlon programs, 180, 183, 200
  mainland production costs of, 270
  price difference for, 113–114
  production of, connected to the refinery, 83–84
Acrylonitrile, as a petrochemical intermediate of the full programs, 69
  mainland production costs of, in Dynel programs, 186, 187–189, 190
  in Orlon programs, 178–179, 181–183
  required for Orlon polymer production, 72–73, 75, 78, 178
  in the Orlon B program, 89
  transport cost of, 112–113
Activities, alternative, prohibition of, in a single program, 67
  minimum scale restrictions for, 66, 68
  of a refinery–petrochemical–synthetic fiber complex in Puerto Rico, 39–50
    basic matrix of, 40–49
    types of, excluded from basic matrix, 50–52
Adipic acid, as an intermediate in nylon salt production, 83n
  transport cost of, 112–113

Adiponitrile, as an intermediate in nylon salt production, 83n
  transport cost of, 111
Advantage, *see* Locational advantage
Air as a commodity with zero price difference, 119
Airov, on the location of synthetic fiber production, 117
  on wage rates in chemical and synthetic fiber activities, 120–121
Ammonia, as a basic petrochemical for the full programs, 69, 79
  in the nylon programs, 159, 164–165, 168, 170
  in the short programs, 201
  production of, connected to the refinery, 83–85
  from fuel oil, as requiring differential profitability correction, 122
  from methane, for Dacron A program, 86
  from refinery gas and fuel oil, for the Orlon B program, 92
  mainland unit costs of, 263–264, 266
  transport cost of, 111
Ammonium nitrate, as a final fertilizer commodity output, 70
  in the nylon programs, 164–165, 170–171, 174
  in the Orlon programs, 178, 183
  in the short programs, 201–202
  mainland production costs of, 268

283

Ammonium nitrate, price difference for purchase of, 119
  for sale of, 118–119
  production of, in the Dacron A program, 86
  transport cost of, 112–113
Ammonium sulfate, as a by-product of adipic acid production, 86n
  transport cost of, 112–113
Aromatics, mainland production costs of, 271

Bagasse, as an input of the full programs, 70
  transport cost of, 112–113
Benzene, as a basic petrochemical for the full programs, 69, 79, 159
  not produced in some Puerto Rico nylon programs, 82
  nylon programs based on, 162n
  production of, at less than minimum scale, 131
    jointly with premium gasoline, 227–229
  transport cost of, 111
Butadiene, as a basic petrochemical for the full programs, 69, 79, 159, 163n
  mainland production costs of, 272
  not produced in Puerto Rico programs, 71, 82
  production of, 229–230
    at the expense of polymer gasoline, 229
  transport cost of, 111
Butane, price difference for sale of, 104
  transport cost of, 111
Butylene, price difference for sale of, 104
  transport cost of, 111

C_2—C_2C_3—C_3, as a mixed refinery gas stream, 80
  composition of, 81, 222–224
  production of ethylene, acetylene, and ammonia from, 82–83
Calcium carbide, mainland production costs of, 272
Capital services, 51–56
  as subject to zero price difference, 52n

Capital services, requirements of, in refinery, petrochemical, and synthetic fiber activities, 54–55
Carbon dioxide, as an input of the full programs, 70
  as a commodity with zero price difference, 119–120
Caustic soda, price difference, for purchase of, 115
  for sale of, 115–116
  transport cost of, 111
Chemical commodities, liquid, transport costs of, 110–112
  price differences for, as equivalent to transport cost, 109–110
  solid, transport costs of, 112–113
Chemical-petroleum labor, price difference for, 121
  as applied in calculation of locational advantage, 129
  reduction of Puerto Rico's disadvantage on, 140
  requirements of, in synthetic fiber activities, 58
  scarcity of, in Puerto Rico, 57, 140
Chilton and "six-tenths rule" applicable to plant cost, 52n
Chlorine, as an input of the full programs, 83
  in nylon programs, 172
  mainland production costs of, 273
  transport cost of, 111
Coke, petroleum, heating value of, 108n
  price difference, for purchase of, 108, 113
  for sale of, 108
  principal U.S. markets for, 108
  transport cost of, 112–113
Coking as a basic refinery process, 217, 221
Commodities, as inputs and outputs of refinery-petrochemical-synthetic fiber activities, 40–50, 30–31
  assumed invariant with respect to location, 50
  total inputs and outputs of, for full programs, 90–91
  types of, omitted from the basic activity matrix, 50–52

INDEX

Commodity flows and the analysis of industrial development in Puerto Rico, 6, 10-11
Comparative cost (industry-by-industry) analysis, as an approach to regional analysis, 6, 7-8
  inadequacy of, in identifying Puerto Rico industrial development, 8
  typical basis of, 7-8
Complex(es), *see* Industrial complex(es); Program(s)
Cost differential(s) for commodities, 95-96
Cost(s) of production, mainland, as utilized in differential-profitability calculations, 146, 149, 153-154, 155
  of acrylonitrile, in Dynel programs, 186, 188-189, 190
    in Orlon programs, 178-179, 181-183
  of basic intermediate commodities, 262-264, 266-282
  of ethylene glycol and dimethyl terephthalate in Dacron programs, 194, 196-197
  of fertilizer commodities, in Dynel programs, 192
    in nylon programs, 164-165
    in Short Programs B and D, 201
  of intermediate commodities in Orlon programs, 178-183
  of nylon salt, 156-157, 162-163, 170, 172-175
  of raw material and intermediate commodities in nylon programs, 156-160
  of vinyl chloride in Dynel programs, 187-189, 191
  requirement of, for petrochemical intermediates, 149
  sources of data for, 263-265
Cracking, fluid catalytic, as a basic refinery process, 217, 221
Crude oil, *see* Petroleum, crude
Cycle oil, as an input for ammonia production, 79
  price difference, for purchase of, 102
    for sale of, 102
Cyclohexane, as an input of the full programs, 156
  not produced in Puerto Rico programs, 71, 83n

Cyclohexane, production of, as an activity excluded from the basic activity matrix, 40-41
  transport cost of, 111
Cyclohexanone, transport cost of, 111

Dacron fiber, as a final commodity output, 69
  production of, in the Dacron A program, 86
  production routes to, 74
Dacron polymer, as a final intermediate in the Dacron A program, 86
  transport cost of, 112-113
Dacron programs, activity scales of, 194-195
  and differential-profitability corrections, 194
  Dacron A, 86-87
    and increased ethylene glycol production, 210
    as most favorable, 195, 200
    compared with Short Program D, 202n, 206
  Dacron B and C, 195
  overall locational advantage of, in Puerto Rico, 198
  reduced Dacron A, 126n, 142, 195, 208-209
  superiority of, over all full programs, 209, 213
Dacron subprograms, 235
Diesel oil, price difference, for purchase of, 102
  for sale of, 102
Differential profitability, contribution to, of integration economies, 150
  corrections for, as affecting preliminary locational advantage, 37, 37n, 138
  as requiring knowledge of mainland prices or costs, 145-146, 148, 155
  as summarized for full and short programs, 212-213
  Dacron programs, 195
  due to petrochemical scale differences, 152-153
  Dynel programs, 192
  economies of scale as a major component of, 138

286 INDEX

Differential profitability, corrections for, full and short programs, assuming continuous-filament production, 203–206
   under differing wage-rate assumptions, 206–208
   necessity of, 37, 37n, 144
   nylon programs, 164–165, 170–171, 175–176
   on the basis of three mainland scale assumptions, 154
   Orlon programs, 183
   short programs, 201–203
   methods of correcting for, 145–146, 153–155
Dimethyl formamide, as a required input of the full programs, 70
   required for Orlon production, 71–72
   transport cost of, 111
Dimethyl terephthalate, as a petrochemical intermediate in the Dacron programs, 86, 192, 194–195
   import of, for reduced Dacron programs, 209
   mainland production costs of, in Dacron programs, 194, 197
   production locations of, in the United States, 109n
   transport cost of, 112–113
Distillation, atmospheric, *see* Topping
Dynel fiber, as a final commodity output, 69
   production routes to, 73
Dynel polymer, transport cost of, 112–113
Dynel programs, activity scales for, 188
   and differential-profitability corrections, 189–192
   Dynel A and B, 192
   Dynel F, as most favorable, 192, 200
   overall locational advantage of, in Puerto Rico, 193
Dynel subprograms, 233–234

Economic development theory, as an approach to regional analysis, 6
   typical concepts of, 7
   unsuitability of, as a technique for identifying specific industrial development of Puerto Rico, 7

Economies of scale, *see* Scale economies
Electronics production, exclusion of, from industrial complexes evaluated for Puerto Rico, 32, 32n
Ethane, as a component of PLAT gas stream, 80
   as a joint product of ethylene separation, 82
   production of ethylene, acetylene, and ammonia from, 83–84, 92
   price difference, for purchase of, 113
   for sale of, 104
Ethylene, as a basic petrochemical for the full programs, 69, 79
   as produced from C_2—C_2C_3—C_3 stream, 80–81
   as produced from pure propane or ethane, 83–84, 92
   in Dynel programs, 186–187
   price difference, for purchase of, 113
   for sale of, 104
   production of, at less than minimum scale, 131
   connected to refinery, 83
   from gas oil fractions, as an activity excluded from the basic activity matrix, 41
   in the Orlon B program, 92
   separation of, as a refinery process, 217, 225–227
   utility requirements of, 225
   unit costs of, mainland, 263–264, 274
Ethylene dichloride, in Dynel programs, 187
   transport cost of, 111
   unit production costs of, mainland, 275
Ethylene glycol, as a petrochemical intermediate in the Dacron A program, 86
   import of, for reduced Dacron programs, 209
   in Dacron programs, 192, 194, 200
   transport cost of, 111
   unit production costs of, mainland, in Dacron programs, 192, 194, 196
Ethylene oxide, as a petrochemical intermediate of the full programs, 69
   in Dynel programs, 187
   in Orlon programs, 75, 78, 180, 183
   mainland production costs of, 276

Fertilizer, as products based on hydrocarbon raw materials, 29
  characteristics of demand for, in Puerto Rico, 32n
  imports of, to Puerto Rico, 29, 65
  price differences, for purchase of, 119
    for sale of, 119
  production of, in a Puerto Rico industrial complex, 5, 32
    limited in Puerto Rico programs, 64
    quantities of, for "standard" program, 85–86
Fiber(s), *see* Synthetic fiber(s)
Fuel, as an input of the full programs, 69
  disadvantage of Puerto Rico with respect to, 140
  price difference for inputs of, based on fuel oil export price, 104
    based on fuel oil import price, 108
Fuel oil, residual, price difference, for purchase of, 102, 108
    for sale of, 102
  price of, in Puerto Rico, as the export price, 103
  production of ammonia from, 79, 83–85
Fuel products, gaseous, price difference, for purchase of, 113–114
    for sale of, 103–104, 108
  price of, compared to liquid fuel, 103
  transport cost of, by pipeline, 103, 103n
Furfural, as a raw material in nylon salt production, 159, 160, 162n
  price difference for purchase and sale of, as equivalent to transport cost, 114–115

Gas, natural, and crude petroleum, as basic raw materials, 28, 28n, 30–31
Gas oil, price difference, for purchase of, 102
  for sale of, 102
Gas separation as a refinery process, 217, 222, 224
Gases, high pressure, price difference, for purchase of, 113
  for sale of, 103, 104, 108
  transport costs of, 113–114
Gasoline, price difference, for purchase of, 102
  for sale of, 102
Gasoline, price difference, for sale of, as applied in the Orlon A program, 126
Gasoline stabilization as a refinery process, 217, 222
Gosfield, and Puerto Rico input-output analysis, 14n
Gravity models, use of, as a technique in the study of industrial development in Puerto Rico, 6, 26
Gulf Coast as a favorable location for chemical production, 109

H_2C_1, as mixed refinery gas, 80
  as used in ammonia production, 80, 83–84
  composition of, 80–81
  production of acetylene from, 80, 83–84
Hexamethylene diamine, transport cost of, 111
Hydrochloric acid, production of, in Puerto Rico programs, 83
  transport cost of, 111
  see also Hydrogen chloride
Hydrogen, as a component of PLAT and H_2C_1 gas streams, 80–81
  electrolytic, as a by-product of chlorine production, 82
  used in ammonia manufacture, 85
  price difference, for purchase of, 113
    for sale of, 104
Hydrogen chloride, in Dynel programs, 187
  in nylon programs, 172
  mainland production costs of, 277
  shipped as hydrochloric acid, 111n
Hydrogen cyanide, as a petrochemical intermediate of the full programs, 69
  in Dynel programs, 187
  in Orlon programs, 75, 78, 180
  in the Nylon A program, 159
  mainland production costs of, 278
  production of, at less than minimum scale, 131
  transport cost of, 111

Ideal weights, use of, in computation of labor coefficient, 11n
Indirect costs, as associated with each category of direct labor, 60

288　INDEX

Indirect costs, as expressed in terms of direct labor and capital requirements, 59–60
Industrial complex(es), and economies of spatial juxtaposition, 33
　definition of, 33
　industrial activities not considered in analysis of, for Puerto Rico, 32–33
　refinery–petrochemical–synthetic fiber–fertilizer, as favorable for Puerto Rico, 28–29, 32
　problems of choice of group of, for comparison, 33–36
　spatial interdependence of activities of, 33
Industrial complex analysis, and oil refining, petrochemical, fertilizer, synthetic fiber production in Puerto Rico, 5
　as an approach to the study of specific industrial development in Puerto Rico, 26
Input coefficients, interregional, changes in, due to changing income-consumption patterns and changes in local and interregional markets, 23–25
　effects of resource limitations on, 23
　expression of, in physical units, $23n$
　stability of, in regions such as Puerto Rico, 22
Input–output analysis, interregional, and the study of economic development in Puerto Rico, 6, 12–25
　choice of appropriate regional sets of industries for, 14
　choice of appropriate set of regions for, 13
　in a hypothetical three-region–Puerto Rico model, 14–21
　interregional input structure of industries required for, 15, $15n$
　limitations and qualifications of, in study of regions such as Puerto Rico, 21–25
Input structure of industries, interregional, as developed for a hypothetical three-region–Puerto Rico input–output model, 15, $15n$, 20–21
　use of, as meaningful economic relationships, 19–21

Integration economies, as contributing to differential profitability, 150
Interindustry relations, need for evaluation of, in regional analysis, 8
Investment cost(s), for nylon salt plants, 156
　less than proportional variation of, with scale, 52–53
Isard and Schooler on transport rates for liquid chemicals, 110, $110n$

Kerosene, price difference, for purchase of, 102
　for sale of, 102

LPG, see Liquefied petroleum gas
Labor, cheap, as a major resource of Puerto Rico, 27
　sites of, as attraction for individual industries, 11–12
　skilled, scarcity of, in Puerto Rico, 27
Labor coefficient, as a technique for analysis of industrial development in Puerto Rico, 6
　as similar to comparative cost computation, 11–12
　limitations to usefulness of, 12
Labor factor(s), and the variation of direct labor requirements with output scale, 56–57
　for refinery, petrochemical, and synthetic fiber activities, 54–55
　use of, to estimate direct labor requirements, 58
Labor services, as divided into two categories, 57
　as inputs, omitted from basic activity matrix, 51
　required in the full programs, 69
　computation of, 93, $93n$
　which do not vary linearly with scale, 51, 56–57
　mainland unit rates for, 263
　price differences for, 121
　requirements of, for refinery, petrochemical, and synthetic fiber activities, 54–55, 94
Lime, as an input of the full programs, 83
　mainland production costs of, 281
　transport cost of, 112–113

INDEX   289

Limestone, as a commodity with zero price difference, 119, 120
  as a raw material input of the full programs, 70, 160
  transport cost of, 112-113
Lindsay, on petroleum tanker transport costs, 96-97, 100
  on refinery processes, 217, 219, 221-222
Linear programming, as an approach to the analysis of industrial development in Puerto Rico, 6, 26
Liquefied petroleum gas, as a basic input for petrochemical production, 79, 83-84
  heating value of, $107n$
  price difference, for purchase of, 111, 113
    for sale of, 107, 108
  recovery of, as a refinery process, 217, 222, 224
  transport cost of, 111
Localization, coefficient of, and the study of industrial development in Puerto Rico, 6, 10-11
Location quotient, and related concepts, as an approach to analysis of Puerto Rico industrial development, 6, 9-10
  limited usefulness of, 9-10
Locational (dis)advantage, as defined for a Puerto Rico complex, 36
  overall, as affected by spatial integration economies, 215
    effect of Puerto Rico tax advantage on, 214-215
    for continuous-filament fiber programs, 203, 205
    for Dacron A with increased ethylene glycol production, 210-211
    for Dacron programs, 195, 198
    for Dynel programs, 189, 192-193
    for full and short programs under differing wage rate assumptions, 206-208
    for nylon programs, 165-167, 171, 175-176
    for Orlon programs, 183-185
    for short programs, 202-204
  preliminary, 37, 127, 135, 141
    adjustment of, to allow for differential profitability, 37, 130-131

Locational (dis)advantage, preliminary, as summarized for full and short programs, 212
  for full programs subject to modification, 138
  for short programs, 132-138, 141
  for short programs compared to full programs, 138-139, 143

Markets, mainland, *see* United States mainland
Methane, as a basic petrochemical for the full programs, 69, 79
  as a component of the PLAT and H_2C_1 gas streams, 80
  price difference, for purchase of, 113
    for sale of, 104
Methanol, as a commodity with zero price difference, 119
  mainland production costs of, 279
  production of, as similar to ammonia production, $69n$
  transport cost of, 111

Naphtha, price difference, for purchase of, 102
  for sale of, 102
New York area as a favorable location for chemical production, 109
New York market for fuel oil, 95
Nitric acid, as an input of the full programs, 83
  in the nylon programs, 164, 170-171
  mainland production costs of, 267
Nitrogen, price differences for, 113-114
Nylon fiber, as a final commodity output, 69
  production routes to, 75-77
Nylon programs, activity output scales for, 158
  differential-profitability corrections for, fertilizer production, 164-165, 171, 175-176
    nylon salt production, 164, 170-171, 175-176
    total, 165, 171, 175-176
  Nylon A, 159, 164-165
  Nylon B, C, E, and F, 174
  Nylon D, 163, 166, 170, 176
  Nylon G, 166, 170, 171, 174, 175

Nylon programs, Nylon G, as most favorable nylon program, 176, 200
  compared with Short Program D, 203n
Nylon H, 166, 170, 171, 174, 176
Nylon J, 159, 167
Nylon L, M, N, and O, 162n
  overall locational advantages of, in Puerto Rico, 166–169
Nylon salt, mainland production costs of, in programs identical with Puerto Rico programs, 157
  maximum scale economies, 162–163
  minimum scale economies, 174–175
  moderate scale economies, 170, 172–173
  transport cost of, 112, 113
Nylon subprograms, 236

Oil, *see* Petroleum
Oil refinery(ies), *see* Petroleum refinery(ies)
Orlon fiber, as a final commodity output, 69, 75
  production of, in the Orlon B program, 89
  production routes to, 72
Orlon polymer, as a final intermediate for Orlon fiber, 71–72, 75
  production of, as an activity in the Orlon B program, 89
  transport cost of, 112–113
Orlon programs, activity scales in, 180–181
  differential-profitability corrections for, 183
  Orlon A, 126, 180
  Orlon B, 89, 180
  Orlon J, as most favorable, 183, 200
    compared with Short Program A, 203n
    production of ammonia requirements for, 85
  overall locational advantage of, in Puerto Rico, 184–185
Orlon subprograms, 78

Paraxylene, as a basic petrochemical of the full programs, 70
  mainland separation costs of, 280

Paraxylene, production of, as an activity excluded from the basic activity matrix, 40–41, 70
  sources of, in the United States, 109n
  transport cost of, 111
Petrochemicals, limitations on production of, in Puerto Rico programs, 66
  primary and intermediate, as products of the full programs, 69
  process differences in production of, 148–149
  differential-profitability correction for, 150
  product differences in production of, 149–150
  production of, as disadvantageous for Puerto Rico, 139–140
  in a Puerto Rico industrial complex, 5
  in Puerto Rico and sale to mainland and Caribbean markets, 210–211
Petroleum, crude, and natural gas, as basic raw materials for diverse intermediate and final products, 28, 28n
  as a basic input of the full programs, 69
  import of, from Venezuela to the United States, 102n
  price difference, for purchase of, 101–102
    for sale of, 102
  price structure of, as determined by the New York market, 100
  reduction of short-haul disadvantage on, 97n
  transport cost of, by tanker, 96–97
  refining of, in a Puerto Rico industrial complex, 5, 29
  Venezuelan, price structure of, 100, 136
    proximity to, as a major resource of Puerto Rico, 28
Petroleum refinery(ies), basic processes of, 217–222
  differences in scale of, 151
  gas flows in, 223, 226
  location pattern of, U.S. mainland, 101
  prototype #4, and relevant process differences, 147
    as a basis for all programs considered, 62–63
    as an activity of Dacron A program, 86

INDEX 291

Petroleum refinery(ies), prototype #4, operation of, restricted to unit scale, 63
refinery gases from, 81
Petroleum refinery prototypes, as activities of basic activity matrix, 45-49
evaluation limited to one, 49n
Plant factor(s), and the variation of investment cost with output scale, 52-53
for refinery, petrochemical, and synthetic fiber activities, 54-55
relationship of, to minimum economic scale, 53n
use of, to estimate plant investment cost, 55-56
"PLAT," as mixed refinery gas, 80
as used for ammonia production, 83, 85
composition of, 222
Polymer commodities, transport of, as necessitating differential-profitability corrections, 122-123, 130
Polymerization, catalytic, as a refinery process, 217, 222, 224-225
Power, as a by-product of steam generation, 105, 106n
as an input of the full programs, 69
price difference for inputs of, based on fuel oil export price, 105, 107
based on fuel oil import price, 109
requirements of, in the Orlon B program, 89
Price difference(s), and calculation of preliminary locational advantage, 37
as applied in Orlon A program calculations, 126
as invariant with amounts sold or purchased, 96
basis for, on refinery, petrochemical, and polymer commodities, 101
for chemical commodities, as equivalent to transport cost, 109
for commodities, purchase of, 95, 98-99
sale of, 98-99
for crude oil, and liquid refinery products, as dependent on transport costs, 96
purchase of, 101-102
sale of, 102
for fertilizer commodities, purchase of, 119

Price difference(s), for fertilizer commodities, sale of, 118-119
for fuel inputs, based on fuel oil export price, 104
based on fuel oil import price, 108
for gaseous fuel products, purchase of, 113
sale of, 103-104, 108
for labor services, 121
as applied in calculation of locational advantage, 129
for liquefied petroleum gas, purchase of, 111
sale of, 107, 108
for liquid refinery products, sale of, 102, 108
for petroleum coke, purchase of, 108
sale of, 108
for power inputs, based on fuel oil export price, 105, 107
based on fuel oil import price, 108
for steam inputs, based on fuel oil export price, 104-105
based on fuel oil import price, 108-109
for synthetic fibers, purchase of, 118
sale of, 116-117
three sets of, as applied to Short Program C, 136
use of a number of different sets of, 37
zero, implications of, 50-51
Process difference(s), as an element of differential profitability, 145, 153
in oil refinery activities, 147-148
in petrochemical activities, 148-150
in synthetic fiber activities, 150-151
in the production of intermediate commodities, 146, 149
interrelations of, with product differences, 146, 149, 155
Product difference(s), abstraction from, in differential-profitability calculations, 146, 149-150, 153
as an element of differential profitability, 145, 153
in petrochemical activities, 152-153
in synthetic fiber activities, 153
Production costs, *see* Costs of production
Production programs, as supplementary to subprograms, 79
Dacron, 248-251

292 INDEX

Production programs, Dynel, 242–247
  nylon, 252–261
  Orlon, 237–241
Programs, alternate, to full programs, 71
  as combinations of refinery, petrochemical, and synthetic fiber activities, 36
  as made up of alternative activities, 62
  continuous-filament fiber, 203, 205–206
  overall locational advantage of, in Puerto Rico, 205
  full, and alternative sets of wage-rate differences, 140–142
    as contrasted to short programs, 133, 140–143
    as showing preliminary locational advantage in Puerto Rico, 127, 138, 141
    calculation of total inputs and outputs for, 88–89
    comparative superiority among, 200
    minor errors in programming and calculations of, 88, 88n
    total commodity inputs and outputs of, 90–91
    total labor requirements of, 94
    violations of minimum scale restrictions in, 88
    with overall locational advantage, 199–200
  general pattern of, 68–71
  objectives in choice of, 68–69
  reduced, 209
  short, and alternative sets of wage-rate differences, 140–142
    as alternative to full programs, 37, 125–126
    as contrasted to full programs, 133, 140–143
    comparisons of, with full programs, 202
    differential-profitability corrections for, 201–202
    individual, overall locational advantage of, 202–204
    preliminary locational advantages of, 134–138, 141
Propane, price difference, for purchase of, 111
  for sale of, 104

Propane, transport cost of, 111
Propylene, price difference, for purchase of, 111
  for sale of, 104
  transport cost of, 111
Puerto Rico, considered as one major region, 13
  principal mainland regions to which tied, 13
  resources of, 27–28
Purchase price of fuel oil as import price, 95

Raw materials, basic, unit rates for, 263
  derivation of costs of, for nylon programs, 159–160
Re-forming, as a basic refinery process, 217, 222
Region(s), as subject to many conceptual classifications, 5–6
  structure of, as a relativistic concept, 6
Regional analysis, validity of several techniques of, 5
Resources, natural, nonhuman, scarcity and low quality of, in Puerto Rico, 27
Revenue differential(s) for commodities, 95

Salt, as a commodity with zero price difference, 119, 120
  as an input of the full programs, 70
  transport cost of, 112–113
Scale economies, as part of differential profitability, 131, 139
  disadvantage with respect to, of Puerto Rico programs, 208
    reduction of, by expanded petrochemical production, 210–211
    by import of intermediate commodities, 208–210
  for mainland Dacron programs, 192, 194
  for mainland Dynel programs, 187–189
  for mainland nylon programs, 162–163, 170, 172–173, 174–175
  for mainland Orlon programs, 179, 182
  in benzene production, 132n
  moderate, as expected to characterize mainland operations, 200–201

INDEX

Scale(s) of output, differences in, and differential profitability, 151
　as types of process and product differences, 151
　in fiber activities, 153
　in oil refinery activities, 151
　in petrochemical activities, 152-153
　for Puerto Rico and mainland Dacron program activities, 192, 194-195
　for Puerto Rico and mainland Dynel program activities, 188-189
　for Puerto Rico and mainland Orlon program activities, 180-181
　for Puerto Rico nylon salt activities, 156, 158, 161, 169-170, 173-174
　increases in, desirability of, 132
　inputs and outputs which vary linearly with, 51
　inputs which do not vary linearly with, omitted from basic activity matrix, 51
　maximum, exceeding of, $131n$
　　for mainland nylon salt activity, 161
　minimum, for mainland nylon program activities, 172
　　for Puerto Rico activities, 68
　moderate, for mainland nylon program activities, 167-168
"Short-cut" method for computing locational advantage, 92, 125, 127-129, $129n$
Short programs, *see* Programs, short
Sodium cyanide, in the Nylon A and J programs, 159-160
　mainland production costs of, 281
　transport cost of, 112-113
Southeastern United States, as location of synthetic fiber activities, 117
Spatial independence, and locational analysis, 34
　of some technologically interrelated activities, 33-34
Spatial integration, economies of, 133, 143
Spatial juxtaposition, economies of, 33, 35-36
　as important to economic development agencies, 36
Split-location patterns, as made possible by spatial independence, 34

Split-location patterns, possibility of, between Puerto Rico and the mainland, 35
Steam, as an input of the full programs, 69
　price difference for inputs of, based on fuel oil export price, 104-105
　based on fuel oil import price, 108-109
Subprograms, as connecting basic petrochemical and final fiber production, 74
　as supplemented by production programs, 79
Sulfur, as an input of the full programs, 70
　price difference for purchase and sale of, 116
　transport cost of, 112-113
Sulfuric acid, as an input of the full programs, 83
　in nylon programs, 172
　mainland production costs of, 282
　price differences for purchase and sale of, 116
　production of, at less than minimum scale, 131-132
　transport cost of, 111
Synthetic fiber(s), as final commodities of the full programs, 69
　as products based on hydrocarbon raw materials, 28-29, $28n$
　continuous filament, overall locational advantage of programs for, in Puerto Rico, 205
　production of, as adding to advantage of Puerto Rico, 64, 203
　programs for, 203-206
　price difference, for purchase of, 118
　process and product differences in production of, 150-151
　production of, in a Puerto Rico industrial complex, 5, 29
　as taking advantage of cheap labor in Puerto Rico, 29
Synthetic fiber activities, restriction applicable to output scale of, 63
　staple and continuous filament, 64

Tetrahydrofuran, transport cost of, 111

Textile finishing activities, exclusion of, from industrial complexes evaluated for Puerto Rico, 33
Textile labor, and a labor factor of unity, 57–58
  as a major input of fiber activities, 71
  low cost of, as a major advantage of Puerto Rico, 133
  price difference for, 121
    as applied in calculation of locational advantage, 129
  supply of, in Puerto Rico, 27, 57
Topping as a basic refinery process, 217, 219–220
Transport cost(s), by tanker, crude petroleum and liquid refinery products, 96–97
  rates of, as varying with distance shipped, 96–97, 97n
  reduction of short-haul disadvantage in, 97n
  on gases, 113–114
  on liquid chemicals, 110–112
  on solid chemicals, 112–113

Unit level operation, 42, 44, 78n
United States mainland, easy access to markets of, as a major resource of Puerto Rico, 28
  regions of, as major markets for Puerto Rico, 13
Urea, and differential-profitability corrections, nylon programs, 164–165
  as a final fertilizer commodity output, 70

Urea, in short programs, 201–202
  in the nylon programs, 164–165, 170
  in the Orlon programs, 178, 183
  mainland production costs of, 269
  price difference, for purchase of, 119
    for sale of, 118–119
  production of, in the Dacron A program, 86
  transport cost of, 112–113
Utilities inputs, mainland unit rates for, 263

Vietorisz, on interregional linear programming, 26n
  on the computation of direct labor requirements, 93n
  on the valuation of fuel inputs and outputs, 109n
  on transport rates for gases in cylinders, 113n
Vinyl chloride, mainland production costs of, in Dynel programs, 187, 191
  transport cost of, 111

Water as a commodity with zero price difference, 119–120
Wessel and the variation of direct labor requirements with output scale in chemical processes, 57n

Xylenes, mixed, yield of, by naphtha reforming and Udex extraction, 70n